Born in Cahir, County Tipperary, Dr Kieran Hickey lectures in geography at the National University of Ireland in Galway. Before that he worked at Maynooth and the Armagh Observatory. His main research interests are the influence of climate change on patterns of storminess, the vulnerability of coastal areas to sea-level rise, and the natural and cultural history of wolves in Ireland.

This book is dedicated to
the memory of my mother and father

FIVE MINUTES TO MIDNIGHT?

Ireland and Climate Change

Kieran Hickey

The White Row Press

IVE MINUTES TO MIDNIGHT?

Ireland and Climate Change

Kieran Hickey

The White Row Press

First published 2008 by
The White Row Press
135 Cumberland Road
Dundonald, Belfast BT16 2BB
Northern Ireland

ISBN 978 1 870132 36 7

Cover: Gerry Watters
Printed by W&G Baird Ltd., Antrim
A catalogue record for this book is available from the British Library

This publication is fully recyclable, biodegradable and uses chlorine free paper from a sustainable forest

View our books at: www.whiterowpress.com

Contents

Preface

This book is concerned with climate change and its effects on Ireland, north and south. It comes out of two drivers in my life. The first is a lifelong interest in climate, climate change, and natural disasters, which I have researched and taught for most of my professional life. This work gave me early notice that something dramatic was happening to the planet's climate, and that we were facing a very uncertain future.

The second driver was an awareness that many people found it difficult to relate to climate change, perhaps because they didn't feel that it touched them personally or locally. This may have something to do with the character of the available reading material. Ireland has an ever-expanding specialist literature on climate change, but so far as I am aware there are few popular works on the subject, and with one exception, no book-length works whatsoever.

In attempting to put this book together, I discovered a possible reason why. Climate change is an enormously complex subject, and many of the issues involved must be radically simplified if they are to have any meaning for the lay reader. Boiling these issues down to their basics was no easy job. Subjects deserving of detailed discussion have had to be covered in a paragraph or even a single line.

This short book sets out the causes of climate change and its implications for Ireland in plain language, with the general reader in mind. It examines what is known about climate change, what we can expect in the future, what level of certainty can be ascribed to these predictions, their likely implications, and where we are not sure, the areas in which more research is needed. I hope it will go some way to clarifying what is going on, and make clear how concerned for our future we really should be.

Finally, I would like to thank the ESB and NIE for their sponsorship of this book. I am also grateful for the support of the Grant-in-Aid-of-Publication scheme of the National University of Ireland, Galway. In addition, I would like to thank everyone who has helped with suggestions and feedback on earlier drafts. These include Neil Reid, manager of Quercus at Queen's University, Belfast, Andrew Cooper, director of the Coastal Research Group at the University of Ulster at Coleraine, and most of all my editor and publisher Peter Carr and White Row Press without which this book would never have been written.

1
What is happening and why should we care?

Climate change: is it really happening?

The simple and straightforward answer is yes, climate change is happening. Until the early 1990s, it was possible for well-informed people to doubt that the earth was growing warmer. While data from weather stations showed that temperature was rising, data collected by satellites between 1979-89 seemed to suggest that the earth's temperature was stable, or cooling slightly. Although the weather station data came from thousands of collection points, and the satellite data came from just a few sources initially, a doubt was sown. Was the earth really warming? When the data sets were examined it was the satellite data which did not stand up to scrutiny. Orbital and timing changes had not been allowed for. When these were factored in, the satellite data and the weather station data told the same story. The earth was warming.

It is not only instrumental data that is giving us this message. The warming trend has been observed, to varying degrees, across the globe. It has manifested itself in retreating glaciers, ice sheet melting, more intense and frequent storms, longer growing seasons, milder winters, earlier phytoplankton blooms in lakes, weakened ocean currents, hotter summers, species change, the drowning of atoll islands, and the coming of thunderstorms to Alaska – a completely new phenomenon. The evidence is overwhelming, and being added to each year. By the dawn of the new millennium, climate change denial had ceased to be credible, and the debate moved on to an equally vexed question: is it us?

EU stands firm on carbon fantasies while Arctic ice grows

ED MILIBAND, our new Energy and Climate Change Secretary, has committed Britain, at this moment of financial meltdown, to an 80 per cent reduction of "carbon emissions" by 2050 – which must go down as the most fatuous utterance ever made by a British Cabinet minister (immediately supported by the Tory shadow spokesman). The only way this goal could be achieved would be to shut down almost the whole of our economy.

A slightly firmer grasp on reality prevails in those countries, led by Poland and Italy, which were last week in Brussels urging the EU to moderate its plans to reduce carbon emissions, on the grounds that this was not the moment to be piling onto Europe's economies costs amounting to trillions of euros. But Gordon Brown, alongside the Commission President, José Manuel Barroso, was at the forefront of those insisting that the EU must stick to its guns.

Brussels's only concession came from the Environment Commissioner, Stavros Dimas, who said he would increase from 35 per cent to just over 50 per cent the amount of "carbon credits" which European industry

would be allowed to b
developing world und
Clean Development M
(CDM). In other word
continue emitting CO
firms would be perm
hundreds of billions
India and elsewher
would be to impose
on those firms, suc
suppliers, which c
operations outside
passed on to their
little or no effect

Just how crazy
already becomi

programme broadcast by the BBC World Service last June (and reported

No, nay, never. Denial is an integral part of the climate change landscape. Sunday Telegraph cutting from October 2008, Newsweek cover July 2007.

Is it us?

The debate about the cause or causes of climate change has involved more than just science. It has touched on politics, business, funding and even issues relating to national sovereignty. At times, indeed, it was hard for science to get a look in. As the debate raged, data accumulated within a wide range of disciplines – climatology, atmospheric physics, glacial studies, paleoecology, environmental and earth science, botany, zoology and many others. As it did, it became ever clearer that, while differing combinations of factors had caused climate change in the past, the prime driver behind current climate change was human activity, specifically the burning of fossil fuels. Greenhouse gases, and particularly carbon, had been produced in such volumes that they had altered the composition of the atmosphere. The issue had passed beyond reasonable doubt.

The sceptics were unimpressed. Legions of not particularly well-informed pundits sniped and scoffed. Sound, empirical studies were dismissed as flawed, pertinent data was deemed irrelevant. Alternative explanations were talked up. The carbon hypothesis was disparaged. The scientists that produced it were denigrated.

The public became thoroughly confused. People didn't know what to believe. Most climate scientists looked on in dismay because they were not wracked by doubt. The scientific consensus (to a level of 99%) was settled and established. The evidence pointed unambiguously towards carbon, and the greenhouse gasses. They were the cause of the problem. Not sunspots, not solar variation, not atomic particles coming from exploded stars, not natural, cyclic climate change, not the chemical action of cosmic rays, not water vapour in the atmosphere, or any of the many other notions that were, and are, in such general circulation.

Environmentalists hit back, arguing that the sceptics were fiddling while Rome burned, that no alternative explanation better fitted the data, that the dissidents relied on studies funded by dubious research foundations, which were fronts for the fossil fuel lobby. I understand this final point. Contrary to some perceptions, it is easier for a climate scientist such as myself to get funding to oppose global warming than it is to research it, such is the financial power of these interests.[1]

Many scientists were concerned by this too. Environmentalists, out of the best of motives, were in danger of overselling climate change, and as a consequence bringing the whole concept into disrepute. They sought to focus political and public opinion on the evidence.

Sifting the evidence

The one thing the climate 'debate' does not lack is evidence. There is an abundance of it. But does it all point in the same direction? We have already touched on the issue of warming. Numerous measurements of different types show that the planet is warming. They demonstrate that warming is not a quirk of 'urban heat islands' or the product of the unrepresentative distribution of weather stations. Nor do temperature rises track population rises, as has also been claimed. The greatest temperature increases have taken place in the world's uninhabited polar regions. The quality of data on key climate change indicators such as temperature, sea-ice, hurricanes, and so forth has also become more robust over time, to the point that anti-warming arguments which might have had some merit a decade ago have little force today.[2]

Ah, say the sceptics. What about 1998, the hottest year on record. If the world is warming, they argue, why hasn't it been hotter since?

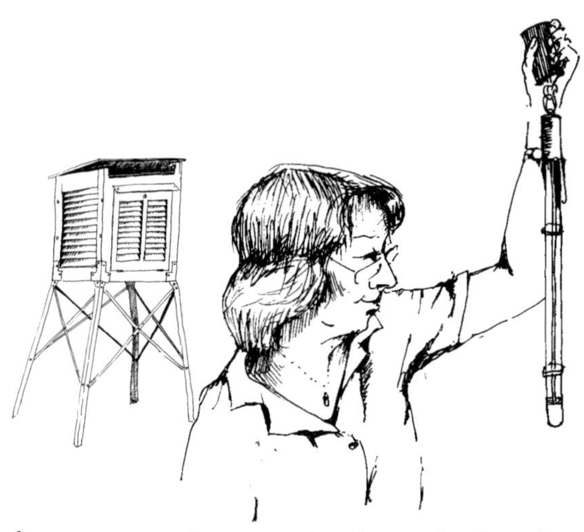

Every day, temperature is measured at thousands of weather stations. (Emma Bayliss)

This seems like a good point. Until we factor in the enigmatic El Niño, a warm sea current which can affect weather patterns across the Pacific (see chapter seven). 1998 broke all records partly because of an exceptionally strong El Niño. That aside, the trends show that warming has continued, both globally and regionally, since 1998.

It is also said, correctly, that there have been periods, millions of years ago, when the earth was warmer than it is at present, the implication being, 'It's happened before, we don't need to change our lifestyles, we'll get through it'. This is factually accurate, but of little relevance to the situation that we face today. The positions of the continents were different, important particulars about the earth's orbit were also different. This simply reminds us that climate change is complex and multi-factorial, and can be caused by different things at different times.

Other claims, such as the idea that the medieval warm period (see chapter three) was warmer than the present, or that the Arctic was warmer in the 1930s than it is today are either untestable or untrue. (The Arctic is currently much warmer than it was in the 1930s.) The issue of ice cores is also raised. Ice core data also shows that sharp temperature shifts have occurred within the last hundred thousand years. These are enigmatic phenomena, and their impact on life will have been swingeing. But, again, they were not a product of the circumstances that we face today. They were caused by a different set of elements in the complex climate equation.

What of the sun? As we shall see in the next chapter, the sun is the most significant determinant of global temperatures, and sunspot activity is an important variable within this. However, cyclical variability in the sun's output accounts for only ten per cent of the level of temperature change caused by greenhouse gases. These gasses account for twenty-five percent of the warming of the planet, clouds add another twenty-five per cent and water vapour the remaining fifty per cent.

This has lead some to identify water vapour as a prime cause of climate change. However, whereas clouds and other forms of water vapour remain in balance with the planet's annual cycle, and do not accumulate, the greenhouse gases stay in the atmosphere for up to hundreds of years, and can play a cumulative role, as they do in the present bout of warming. Water vapour is properly accounted for in climate models, which incorporate it as a feedback.

Modelling is another battleground. We rely on computer models, called Global Circulation Models, for our predictions of future climate, and a huge effort has been put into making these statistically reliable. But are they any good? Critics call them glorified crystal

US cartoon implying that acceptance of global warming is 'un-American'.

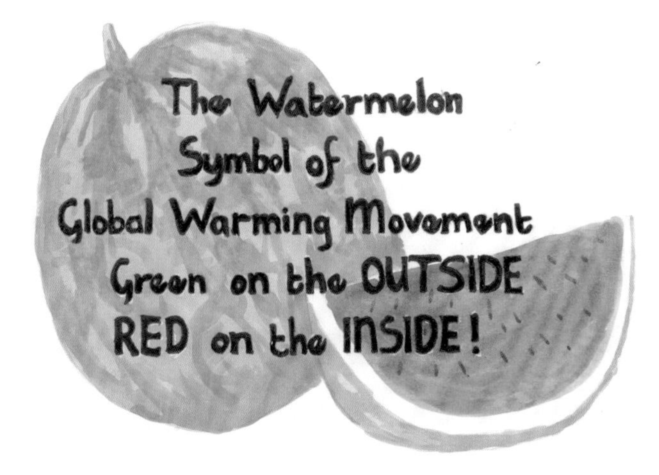

balls, but this does them a disservice. They make extremely useful tools, and are as close as we are likely to get in being able to predict something as complex and multi-causal as climate.

The power and competence of these models has increased enormously over the last decade. Each year, new data has been added, methodological improvements have been made, and more computing power has been brought to bear, making the models increasingly sophisticated and authoritative. They don't quite make the future visible, but they do cast a strong beam into the dark.

What this means for each of us

Why should we care? If we cannot contain the momentum of climate change, then our future looks less than rosy. The transformation will not be noticeable on a daily or even an annual basis, but the world that our children and their children inhabit will be a much altered one.

There are three main reasons, as I see it, to care about this. Firstly, it is in our own interests to adapt to change and make the best of things. Secondly, as the burden of responsibility for global warming is in our hands, so the remedy, or at least the finger that is put in the

'The transformation will not be noticeable on a daily... basis, but the world that our children and their children inhabit will be a much altered one.'

dyke must be ours also. And thirdly, if we want this planet to remain a comfortable home for humankind, then we need to care enough to take immediate action, individually and collectively.

This means changing the way we live our lives. Already, many people have started to make positive changes. They are beginning to take more control of their energy and resource consumption, with a view to becoming more energy efficient and cutting down on waste.

It's also about mindset, and making lifestyle changes such as recycling our waste, insulating our houses, using public transport more often, and even growing some of our own food. But it goes further. We should also think about seeking to influence the people that govern us, the companies that make our machines, provide our services, and so on. For as voters, as consumers, as investors, or even as members of the local golf club or parent-teachers association, we have a little bit of influence, and we should use it.

What this means for government

Government in both parts of Ireland must think strategically if it is to come to terms with the implications of climate change. Both jurisdictions face similar issues. First, there is the need to meet carbon dioxide and other greenhouse gas emissions targets. This will involve – already has involved – significant changes in energy and resource production and consumption, including a much stronger emphasis on drawing energy from renewable sources.

Secondly, there is a need to protect high value, including urban, coastlines from sea level rise. We will also need to meet the increased demand for water and adjust to the increased seasonality of rainfall (see chapter six). Agriculture in particular will need support if it is to successfully adjust to the new conditions. Major changes lie ahead. Meeting these challenges will involve government funding, running into billions of pounds or euros for each territory.

However, the difficulties we face should be put in perspective. A recent report ranked Ireland the second least vulnerable country in the world to the implications of climate change, after Canada. The United Kingdom, including Northern Ireland, was ranked twelfth. So the impacts we face, though serious, are not on a par with those faced by nations elsewhere. The lowest ranked countries were Haiti, Rwanda, Chad, Ethiopia and Afghanistan, which are amongst the poorest and most politically and militarily unstable countries in the world.[3]

Global implications – famine, refugee movements and war
Although the island of Ireland is relatively well placed to meet the challenges of climate change, we do not live in isolation. What happens elsewhere will affect us too. Rising temperatures will make places like the Sahel in Africa virtually uninhabitable. Other regions, including large parts of Bangladesh, will disappear under water. Some islands in the Maldives, and many idyllic-looking Pacific coral islands – the stuff of tourism brochures – have already gone.[4]

Basic resources, including food and water, will become increasingly difficult to obtain. Many regions will find it hard to sustain their existing populations. Possible consequences include acute social tensions, famine, wars over resources, and mass movement of peoples on a scale not seen for centuries, with the UK and Irish Republic becoming potential destinations for refugees. Climate change makes our collective future look increasingly uncertain. The notion that it isn't happening is like the claim that ostriches bury their heads in the sand. They don't, of course. If they did they would suffocate. We should not do the same.

2
What causes climate change?

There is nothing new about climate change. The earth's climate has been in transition since the formation of the planet, thanks to the action of a wide variety of natural factors operating over different timescales to differing and sometimes contrary effect. This chapter identifies the key natural drivers of climate change: solar variation and sunspot cycles, the earth's orbital variations, carbon dioxide levels, volcanoes, and meteors and comets. It then looks at the role played by humans in the changes that we are experiencing today.

The role of the sun

The sun, the font of all life on earth, is the prime determinant of our climate. For centuries, the assumption was that the sun's output was constant. This assumption was empirically unchallengeable, because atmospheric distortion made it more or less impossible to reliably measure the sun's energy. The breakthrough came with the advent of satellites, which made it possible to monitor solar radiation accurately enough to identify changes.

Although it is only in its infancy, satellite monitoring has already delivered important insights. It has allowed us to quantify the output of the sun, at 1,368 watts per metre squared, and shown us that variations of 0.1-0.3% occur from day to day, week to week, and so on. Scientists strongly suspect that greater fluctuations in solar output are likely to occur over longer timescales. But as yet we don't know this, and that is hardly surprising. The satellite record only goes back some forty years, a minute fraction of the lifecycle of the sun.

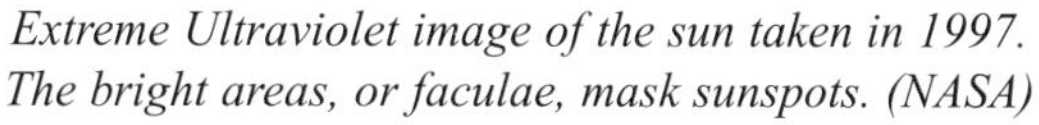

Extreme Ultraviolet image of the sun taken in 1997.
The bright areas, or faculae, mask sunspots. (NASA)

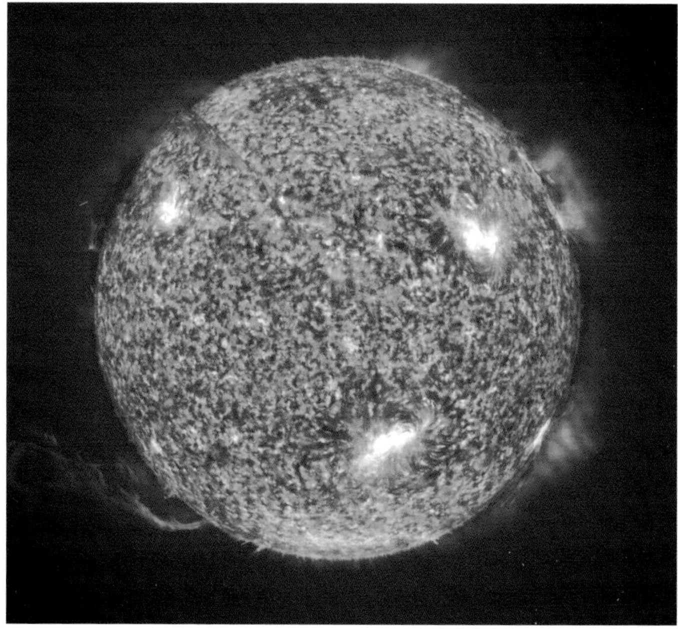

The aspect of the sun's behaviour that is of most relevance to climate change is the occurrence of sunspots. These are large storm-like disturbances on the surface of the sun, which when viewed appear as darker patches. Large sunspots can be several times the size of the earth. Sunspot activity has been recorded for almost a thousand years, with some of the earliest recordings having been made by Chinese astronomers.

Sunspots are also relatively *cool*, with a temperature of around 4,500 degrees Kelvin in comparison to the much warmer 5,800 degrees Kelvin of the rest of the sun's surface. What is most extraordinary and paradoxical about them, however, is that though they are areas of *lower* temperature, when they occur in large numbers they are associated with *increased* solar output, and vice versa. Scientists have, as yet, no clear understanding of why this should be the case.

Sunspots also have a remarkable regularity. They tend to appear, multiply, and disappear again in an eleven year cycle. This eleven year cycle has been linked to changes in solar output, which means that we can use sunspot numbers to track long term changes in the sun's energy levels.

The last four hundred years have witnessed significant annual variations in sunspot numbers. Figure 2.1 shows the eleven year cycle of peaks and troughs in terms of sunspot numbers. It is also clear that this cycle occasionally breaks down or becomes much less pronounced. Two minimums have been identified over the last four hundred years. The largest and most obvious is known as the Maunder Minimum, which occurred over the period 1645-1715, when virtually no sunspots were recorded. Significantly, these years coincided with the coldest part of the Little Ice Age, which will be described in more detail in the next chapter.

A dramatic recovery in sunspot numbers followed over the period to 1790, when the second and much smaller minimum is recorded. Known as the Dalton Minimum, it lasted from 1790-1820, another colder spell in the Little Ice Age. Some – but very little – sunspot activity was observed during this period. From 1820, the eleven year cycle becomes much more pronounced, with very large numbers of sunspots being recorded, so much so, that the period from 1820 onward has been designated the Modern Maximum. Sunspot activity has been especially pronounced since

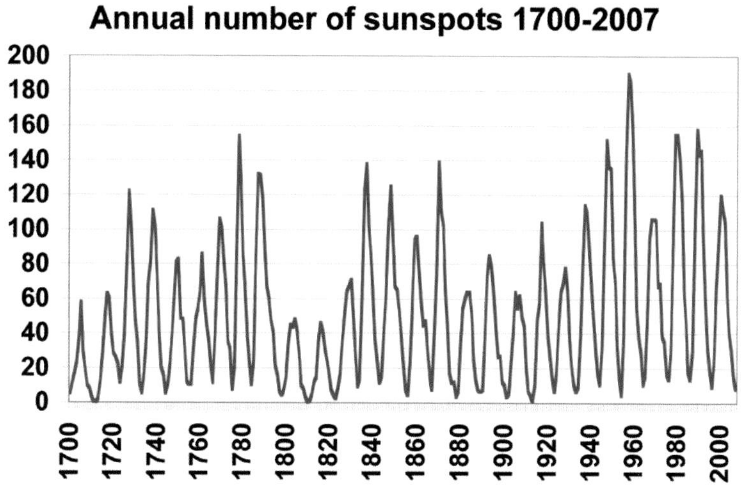

Annual number of sunspots 1700-2007

Sunspot cycles have a dependable, eleven-year rhythm.

1950. This has contributed to, but has not been the main driver behind, the warming of the earth.

At the moment of writing, we are between sunspot cycles. 2008 sees the end of solar cycle twenty-four and the start of solar cycle twenty-five, so few sunspots are being recorded at present. The new cycle is expected to peak in late 2011-12 or early 2013, no two cycles being exactly alike. This means that we are quite likely to see new global temperature records being set in the next few years.

Solar radiation is the explanation of modern climate change favoured by those who do not accept that CO_2 levels play a role in global warming. It is easy to see the attraction of this argument. Solar radiation contributes to the current warming of the earth. But it is not the sole or primary cause, as sceptics have argued. There has been no positive trend in any solar index since the 1960s, meaning that solar forcing cannot be responsible for the recent temperature rise.

Orbital variations

Just as the climate is always changing, the earth itself is in constant motion, as is our solar system and the galaxy that it is a part of, the Milky Way. Despite the fact that most galaxies are moving away from each other, the Milky Way is on a collision course with the galaxy known as Andromeda, a cataclysmic event that is likely to occur in around three billion years or so. The Milky Way is a rotating, spiral galaxy. Our solar system also rotates, travelling at approximately 250 kilometres per second, which means that it completes an orbit of the galaxy every 200-250 million years.

Given these competing influences, it is not surprising that the earth's orbit around the sun is constantly changing. These complex changes are known as orbital variations. Three have been identified as being particularly relevant to climate change. Milutin Milankovitch (1879-1958), a gifted Serbian mathematician, usually receives the credit for this discovery, although others had aired some of these ideas before him, and the Milankovitch cycles, as they are known, identify and describe the three most important orbital variations.

The first variation is in the shape of the orbit of the earth around the sun. Our orbit rarely follows a perfect circle and is most often elliptical or oval-shaped. The extent of the ellipse varies. It is currently running at around three per cent and will reach about five per cent when our orbit is at its most elongated. This orbital irregularity completes its cycle every 100,000 or so years. It is a major determinant of climate, and its action has been linked with the recurring ice ages that the earth has experienced over the last 1.6 million years.

You might think that five per cent isn't much of an ellipse, and you'd be right. But consider the huge climatic impact that this seemingly trifling difference has had. The idea of small differences having massive consequences will recur again and again in our analysis, and it's absolutely central. Were we to look for a sporting parallel, we might cite the men's hundred metres in Beijing. Richard Thompson of Trinidad and Tobago ran 9.89 seconds, and won a silver medal. Craig Pickering ran 10.18, and was eliminated in the qualifiers. The difference between Olympic glory and failure is small.

So it is with the earth's ecosystems. Though wonderfully robust, most of the planet's geophysical and eco-systems work within relatively narrow tolerances. And it doesn't take a whole lot to knock them off their stride. Understand this, and you are well on your way to understanding the dynamics of climate change.

But back to Milankovitch. The second variation the Serbian genius identified was in the tilt of the Earth's axis. Milankovitch discovered

Orbital variations

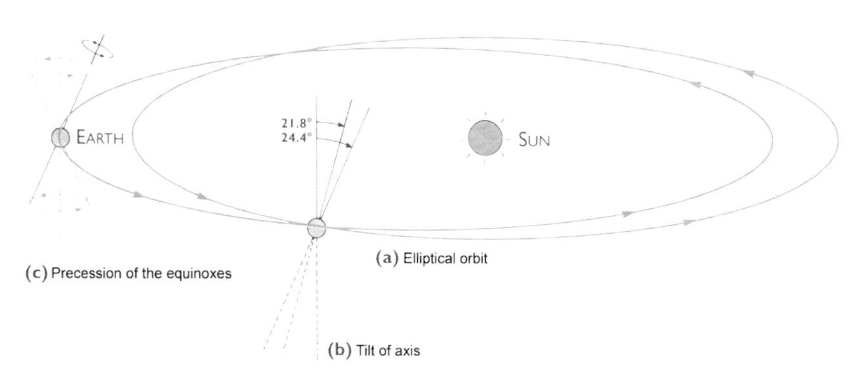

Orbital variations. The inner and outer rings loosely indicate the extent to which the earth's orbit varies. (Gary Hincks)

that this 'axial tilt' was not stable, and ranged from 21.8-24.4 degrees across its 43,000 year cycle. It currently stands at 23.5 degrees. The axial tilt gives us our seasons, and its extent influences their mildness or severity. When there is less axial tilt, the sun's warmth is more evenly distributed between winter and summer, but the difference in the amount of heat received by the tropics and the polar regions also becomes more pronounced. When there is more axial tilt, the opposite is true. In this circumstance the sun's warmth is less evenly distributed between winter and summer, and the difference in the amount of heat received by the tropics and the polar regions becomes less pronounced.

The third variation is called the precession of the equinoxes, and this relates to the way the earth gyrates or 'wobbles' as it spins. As the planet turns, the gravitational forces exerted by the sun and moon cause the earth's axis to trace a conical shape, much like that of a spinning top as it starts to slow down. This orbital variation has a cycle of 26,000 years. Put simply, precession is concerned with where the seasons occur in relation to the elliptical path of the earth's orbit, and its effect is to make the seasons either more or less extreme.

At present, precession is a force for moderation in the northern hemisphere. The earth's orbit brings it closest to the sun in January, when precession causes its axis to lean away from the sun. In other words, when the northern hemisphere experiences winter and receives the *least* amount of sunlight, the earth as a whole receives the *most*. This makes northern winters milder and northern summers slightly cooler and the southern hemisphere's seasons more extreme. In 10,500 years' time the position will be reversed, with the northern hemisphere's summers occurring when the earth is closest to the sun, meaning that summer will be slightly warmer than usual and winter slightly cooler.

The interplay of these three variables has controlled the length and severity of our ice ages and interglacial warm periods. Paradoxically, however, these orbital variations would have little impact were it not for the asymmetric distribution of the earth's landmasses. Landmass is not evenly distributed across the globe. At present, most of the earth's land lies in the northern hemisphere, and is drifting slowly northwards. This makes it easier for ice to form on these lands and their mountain ranges, which in turn cools the climate around them. Had our landmasses happened to lie mostly on the equator, then more warming could have been expected. Likewise, were the earth in the middle of an orogeny, or period of mountain building, the effect of the variations would be cooling, and so on.

So the orbital variations are not only influential *per se*. Their influence is affected and greatly magnified by their interaction with terrestrial variables, a timely reminder of how complex and multi-factorial climate is.

The importance of carbon dioxide

The earth's atmosphere plays a crucial role in controlling the temperature and climate.

It has been a long time in the making. The first evidence of the development of an atmosphere comes around 4.2 billion years ago, but there is no evidence of free oxygen in the atmosphere until 3.4 billion years ago. Around 2.4 billion years ago, an oxygen rich atmosphere began to form. It developed slowly; by 545 million years ago oxygen levels were still only running at about 18% of present day values.

After this period, we begin to see increasing amounts of carbon dioxide (CO_2 or carbon, for short) in the atmosphere, and associated with it, a very warm climate. Temperature and carbon levels would appear to be inextricably linked, and we find that as levels of carbon

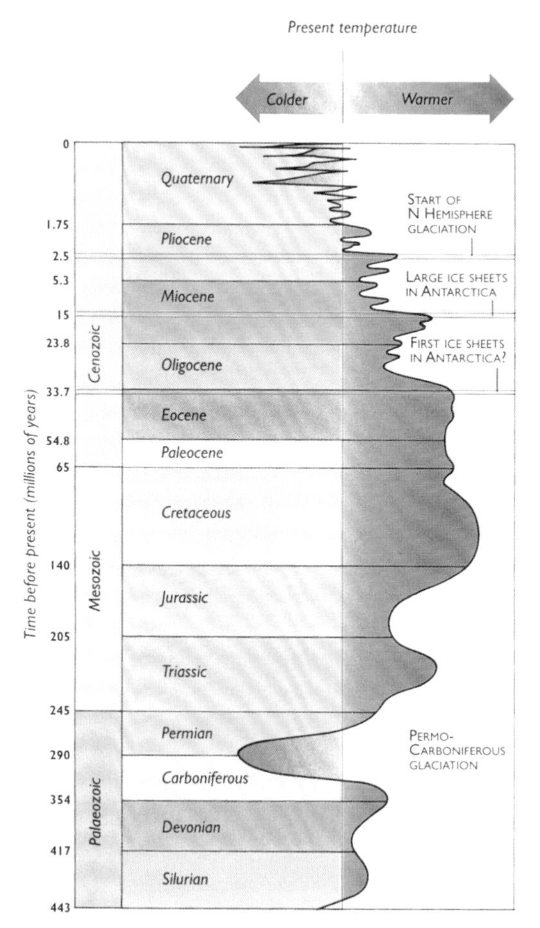

Relative global temperatures, from 443 million years ago to the present. (Michael Eaton)

dioxide continued to rise, up until about 460 million years ago, the climate continued to warm. Then carbon dioxide levels declined. Temperatures fell with them, to the point that, 443 million years ago, ice masses had begun to form at the poles.[1]

The link between atmospheric carbon levels and temperature is complex, but well established. Historically, temperature rises have often preceded and then developed in conjunction with carbon rises. This is because the temperature rise prompts the slow release of large amounts of stored carbon from the world's oceans, which then contributes to further temperature rises – very different circumstances to those we face today.

Broadly speaking, the more carbon dioxide in the atmosphere the higher the temperature, unless some other factor intervenes. During the Devonian period (417-354 million years ago), for example, temperatures rose, not withstanding a dramatic fall in carbon dioxide levels. The fall in carbon was due to the explosion of plant life. Trees eat carbon. Their leaves draw it from the air, and they use it to build their wood, bark and leaf tissues (were one to be brutally unromantic, one might describe them as tubes of crystallised carbon). But trees are more than passive consumers of carbon dioxide, they convert it to oxygen and energy in a process known as photosynthesis, and as such are a dynamic part of the carbon equation.

Carbon dioxide levels reached their nadir around 300 million years ago, recovered during the Mesozoic period, around 180 million years ago, and then declined down to the point at which humans started burning fossil fuels, thus raising carbon dioxide levels once again. In spite of these generally falling carbon dioxide levels, the climate remained warm until about twelve million years ago, when cooling started to set in as carbon dioxide levels became very low. Our own period, the Quaternary, with its regular ice ages and modestly warmer interglacial periods, is also associated with very low carbon dioxide levels.[2]

As this brief atmospheric history shows, when carbon dioxide levels rise, the climate tends to get warmer because of its ability to trap incoming solar radiation or heat from the sun. This is known as the greenhouse effect, and it is offset by the presence of plants, which convert vast quantities of carbon dioxide to oxygen. The interplay between the two has kept temperatures broadly tolerable for aeons.

But now there is a new ingredient in the mix – man. Human beings have raised carbon dioxide levels while at the same time clearing hundreds of millions of hectares of forest and other plant life. Some six million hectares of forest are cut down annually, according to the United Nations. Were we to liken what we have done to, say, running a bath, we have turned up the hot tap at the same time as we have turned down the cold. In the circumstances, it is small wonder that we are getting burned.

Phenomena that affect climate

1st Order	Potential temperature change
Solar luminosity	20-40°C
Solar system geometry	20-40°C
The earth's atmosphere	20-40°C
2nd Order	
Orbital variability	5-15°C
Solar variability	5-15°C
3rd Order – short, sharp shocks	
Volcanoes	0-5°C
Sunspots	0-5°C
Meteorite impacts	0-5°C
El Niño, La Niña	0-5°C

The role of volcanoes

Volcanic eruptions can play havoc with global weather. The eruptions of Tambora in 1815, Krakatoa in 1883, and Pinatubu in the Philippines, which erupted in 1991, are excellent examples of this phenomenon.

Large volcanic eruptions have a marked, but transient effect on climate.

However, contrary to the impression given by Hollywood disaster movies, it is not the emptying of millions of tons of dust and ash into the atmosphere that does the most damage. It is the release of chemicals known as aerosols, particularly sulphur. These are hurled into the upper atmosphere, and then carried across the globe by the wind, an effect that occurs all the more easily if the eruption occurs between the equator and the tropics.[3]

These higher concentrations of aerosols scatter incoming solar radiation, reducing the amount that reaches the surface of the earth, which then cools. A global temperature drop of 0.2 to 0.5°C can occur as a result of a major explosive volcanic eruption, and the effect can last from one to three years, fading as the aerosols are rained out of the atmosphere.

Tambora is a 'classic' example of this kind of disaster. The aerosol barrier cut out heat, making 1816 an exceptionally cold year, characterised by poor harvests and rapidly rising food prices all across the planet.[4] In Ireland, the crisis produced food shortages and an agricultural depression. This led to a general failure of the rents that ensured that the crisis hit every grouping in society, with even the Londonderrys, one of the country's richest landowning families, being reduced to living parsimoniously with few servants.[5]

1816 became known in Irish folklore as 'the year without a summer'. The meteorological data collected at the Armagh Observatory sets out why. During the first quarter of the nineteenth century, mean spring temperatures were lowest between 1816-18. Cloud cover was greatest for the same three years. Rainfall was highest for 1816, and the months with most snow were March 1817 and March 1818 – a catalogue of misery that is unlikely to be due to chance alone.[6]

The role of meteors and comets

Meteors are lumps of rock, or rock-and-ice in the case of comets, which hurtle through space. When the larger ones (objects with diameters greater than about a kilometre) strike the earth, as they do every hundred thousand years or so, they can have a powerful, but short-lived effect on climate.

Comets which cross the earth's path act in the same way as large volcanoes, in that they fill the atmosphere with debris which reduces incoming solar radiation, causing a temporary downturn in climate.

But these are also relatively rare phenomena. Perhaps the last comet to alter our climate was Biela, in c.400-600AD.

The role of humans

Before the Industrial Revolution, human beings made little impact on the atmosphere. We affected it indirectly, through land use changes, when we moved from hunting and gathering to farming. By turning forests into farms, we increased the amount of incoming solar radiation that was reflected back into the atmosphere, causing it to slowly warm.

Since the Industrial Revolution, we have been affecting it more directly, as we have burned increasing amounts of coal, and latterly, oil and gas. (Eighty per cent of man-made carbon dioxide is generated by the burning of these three fuels.) This has released vast amounts of gas into the atmosphere, changing its composition in subtle but significant ways.[7]

The most important of these gases is carbon dioxide. In 1750, CO_2 levels in the atmosphere were 280 parts per million (ppm). By 2008, they had risen to 387 ppm. They are currently rising at a rate of

Smoke stack. Since the Industrial Revolution, we have burned increasing amounts of coal, raising the levels of carbon dioxide in the atmosphere.

0.53% per annum, a trend that will not be easily or painlessly reversed. Were all coal burning to cease by 2020, it would take two hundred years for carbon dioxide levels to fall to 350 ppm, a level generally deemed to be sustainable.

The next most important gas is methane (CH_4), which, in terms of temperature rise, is twenty-three times more potent than carbon dioxide. This has risen from 700 parts per billion (ppb) in 1750 to 1700 ppb by 2006. At present, methane levels seem relatively stable, but it is not clear why this is the case. Few scientists believe that this 'stability' is more than temporary. Most view it as a possible 'pause for breath', before a period of renewed rise, as the conditions underlying the overall increase have not greatly changed.

The third most important gas is nitrous oxide (N_2O). This increased from 270 ppb to 320 ppb between 1750-2006, and is currently rising at the rate 0.25% per annum. The next most important group of gases are the chlorofluorocarbons (CFCs), which most people know as the propellants in aerosol cans and the coolant in fridges and air conditioning units. These did not exist in 1750, and were invented during or after the1920's. CFCs destroy upper atmosphere ozone molecules. They do this with terrifying efficiency, each individual CFC molecule, depending upon its type, being capable of destroying tens of thousands of ozone molecules.

By the 1990s, a mere blink of an eye in climatic terms, CFCs had blasted a hole in the earth's ozone layer, allowing deadly forms of solar radiation through to the earth's surface. This danger, once identified, was well managed globally via the Montreal Protocol, which outlawed the most damaging CFCs. The atmosphere responded quickly. The ozone hole stabilised.

However, the increasing abundance of greenhouse gases has had more insidious effects. They have made the atmosphere absorb more radiation and heat up. There is nothing intrinsically bad about this. The atmosphere is the planet's insulating blanket. Without it, our temperatures would be some thirty-three degrees centigrade lower than they are. So, as ever, it is a question of degree, and of living in a way that does not upset the checks and balances that our ecosystems depend on. Unfortunately, we are not doing this. We are moving beyond these tolerances, with serious implications for all life on earth.

3

The backstory – climate change over the last two thousand years

Our climate is not fixed, it is constantly changing, which means that the phenomenon we call 'climate change' is not the exception, it is the rule. And one does not need to go back to the Pleistocene to find climate change. Ireland has experienced some remarkable lurches in climate over the last two thousand years. This chapter tracks these changes using Irish, British, European, and other data, as it is not at this point possible to construct a detailed history of Irish climate change from Irish sources alone. The next four chapters will delve more deeply into current climate concerns.

The first centuries AD

These centuries were dominated by mild weather, and abundant (but not excessive) rains, which made ideal conditions for good harvests. Temperatures were probably similar to those of the mid-twentieth century. They may even have reached current levels, as evidenced by the almost continuous retreat of glaciers in the Alps and the opening

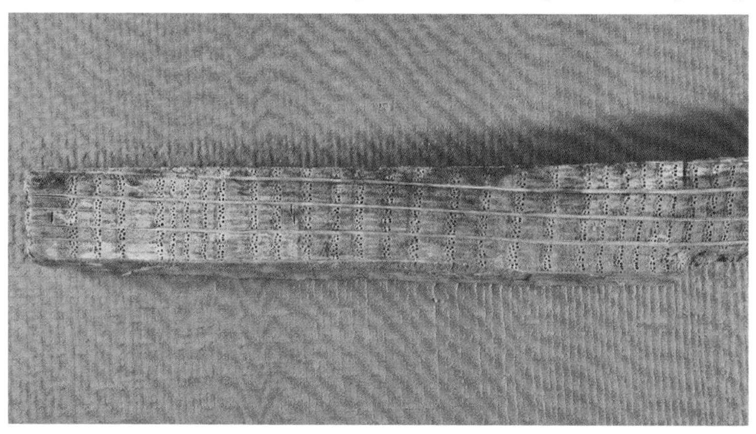

The Belfast Tree Ring Chronology allows events which influenced Ireland's climate to be dated to the year. The annual growth rings can be seen as horizontal lines on this piece of oak from Killarney. (David Brown, QUB)

up of new Alpine passes. Previously uncultivated uplands were farmed. Vines were grown in many parts of England.

This long, mild spell peaked around 200AD, after which temperatures gradually declined. Around 300AD, a sharper downturn began, intensifying between 400-450AD. This change in weather patterns went hand-in-hand with significant alterations in rainfall. There was too little rain in southern Europe and too much rainfall in the north, leading to regular crop failures, famine conditions, and the return of long, snowy winters, which north-eastern Europe had been spared for hundreds of years.

The supply of food also became far less reliable. The farming of upland areas was abandoned and the deteriorating conditions encouraged large northern tribes to move south in search of less inhospitable lands to farm; migrations that played a large part in the disintegration of the Roman Empire.

The Irish data, which is drawn from peat cores, pollen records, and oak tree ring chronologies, broadly reflects this pattern of change. The Belfast Tree Ring Chronology allows seminal events to be dated to the year, and offers us an insight into the general climate. The Irish Annals make just ten references to weather events in pre-Christian Ireland. Two are wonderfully illustrative of the change from benign to more testing climatic conditions. In 15AD it is recorded that:

> prosperous weather was in Ireland during this time. For these were tranquil honourable times, fertility in the fruits of the earth, fish in the river, abundance of milk from the cows, and the tree tops were drooping with the weight of fruit…

The year 436AD, however, was remarkable for a ‘huge snow’.[1] Things had taken a turn for the worse.

The Early Christian period c.450-800

Whatever good the arrival of Christianity did for Ireland, it did not do much for the weather. St. Patrick left the stone unturned. Ireland’s climate continued to deteriorate. This period is associated with poor temperatures and frequent late and early frosts which ruined crops, creating chronic food shortages and famine across Europe.

The 530s-540s were a particularly harsh time. In 542 the bubonic plague reached Europe, killing millions. Another, equally cataclysmic natural event occurred in 535-36, when, according to Michael the Syrian:

> the sun became dark and its darkness lasted for eighteen months. Each day it shone for about four hours, and still the light was only a feeble shadow… the fruits did not ripen and the wine tasted like sour grapes.[2]

There are various explanations of this darkness, the most likely being that it was caused by a massive volcanic eruption, probably of Krakatoa. It lies between Java and Sumatra and, as we have seen, erupted again in 1883 to devastating effect. The eruption of 535, however, appears to have been even more massive, as its climatic effects were harsher and more enduring. It threw such a quantity of aerosols, dust, and ash into the upper atmosphere that it virtually blotted out the sun for a year and a half, causing a dramatic cooling of the earth's surface, and yet another agricultural crisis.

Irish tree ring data suggests that eruptions occurred in 535 and again in 540 or 541, as the years following these dates show a remarkable collapse in tree ring growth.[3] The crisis is vividly represented in the Irish Annals, which note 'blood' (probably volcanic dust mixed with rainfall) falling from the sky in 535, and

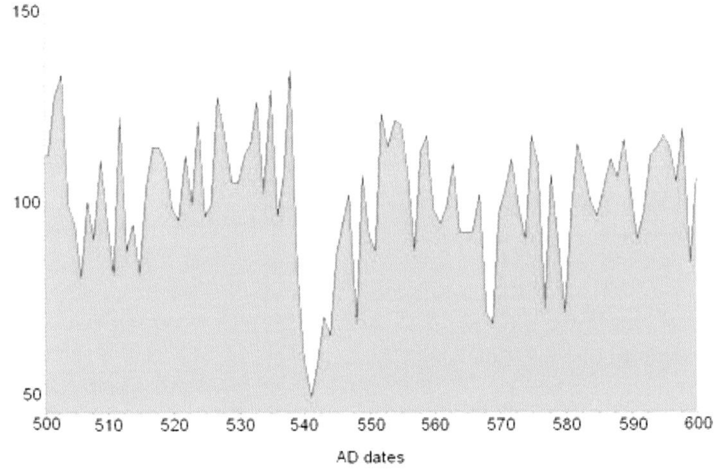

The climatic crisis of the early 540s as reflected in Irish tree ring growth, which more than halved in 540-41. (David Brown, QUB)

failures of bread throughout the years 536-39. This lead to many deaths. In 540, as the Earth pulled out of crisis, a spectacular aurora was observed, created by the diffraction effect of volcanic aerosols in the atmosphere.[4]

However, we should not allow ourselves to be distracted by dramatic, arbitrary events. We must look beyond them and ask, what produced the broad, centuries-long climatic downturn within which they occurred? What made it bite so deep? Was the cyclical climatic downturn extended and intensified by extra-planetary activity? It may have been. Between 400-600 the earth is believed to have experienced a sustained bombardment by meteor showers associated with the break up of the comet Biela.[5] Whatever the causes of the downturn, by 600 the worst was over, and between c.600-800 a slow recovery began.

The medieval warm period 800-1300

After c.800, temperatures returned to levels that were comparable to those of the mid-twentieth century or even today. There was a general retreat of glaciers, sea ice and ice sheets. There was less rain, leading to recurring droughts and water shortages. But these did not derail the general return to prosperity, and the population grew.

Across Europe, farmers re-colonised the uplands, and in England flourishing and extensive vineyards were to be found as far north as York. Abundant harvests are regularly recorded throughout Europe. Society moved beyond subsistence. Labour could be spared to build the magnificent cathedrals, castles and monasteries that are counted amongst the glories of the High Middle Ages.

One of the most striking consequences of medieval climate change is the remarkable ninth century expansion of the Vikings from their homeland in Scandinavia into the British Isles and north-western France, then westwards across the Atlantic to North America. This was facilitated by the northward retreat of the polar ice cap, the gradual disappearance of pack-ice and icebergs from the sea lanes, and a marked abatement in the storminess of the North Atlantic. These changes allowed the longboats to range more freely. It also gave them a longer sailing season, removing one of the principal barriers to exploration.

By 860 the Vikings had begun to explore Iceland (although Irish monks can stake a strong claim to have found it first, thanks to an

Relic of the medieval building boom. The twelfth century ruins of Inch Abbey, County Down.

earlier break in the weather!) Aided by the continuing retreat of Atlantic sea ice, which made for stable lines of communication, and ongoing improvements in temperature, which made farming more viable, the population of Iceland rose to a healthy 77,000 by 1095.

In 982, encouraged by this success, Erik the Red founded a settlement on the southern tip of Greenland. This also thrived, and by 1200 its population had risen to 6,000. Eventually the Vikings made it as far as North America, settling at L'Anse aux Meadows on the northern tip of Newfoundland.

The Irish Annals confirm this generally rosy climatic picture. They suggest that Ireland experienced heat waves, droughts and summer lightning. In 1095, for example, it is recorded that, 'this year was the year of the heat, so that there is no reckoning the number of people whom it killed'.

In the year 1129 there was 'a torrid summer… in which the streams of Ireland dried up'. In 1191 it is stated that 'the river of Galway the Corrib was dried up for several days, so that all things lost in it from time immemorial were recovered and great quantities of fish were taken by the inhabitants'.[6]

The cold returns 1300-1450

But these days of plenty were numbered. After around 1300 the climate again worsened. There was a gradual drift back to colder and wetter conditions. The glaciers and ice sheets re-advanced, and there was a southward movement of sea ice. The Viking colonisations of Iceland and Greenland came under such severe pressure that mass evacuations back to Denmark were considered.

To add to the colonies' problems, the evidence from the Greenland ice sheet suggests that in about 1420 the North Atlantic became much stormier.[7] All over Europe, farmers retreated from upland areas. Climatic stresses produced social stresses. There were mass migrations southwards. Outbreaks of famine and plague almost halved the population of Britain.

In Ireland, the Annals track the deterioration through references to famines caused by lower temperatures and unbearable rainfall levels. Two particularly horrific periods stand out, 1315-17 and 1348-49, when climate-derived famines caused the deaths of several hundred thousand people.

The Little Ice Age 1450-1850

Between 1450-1850, the 'Little Ice Age' gripped Europe. Things were at their bleakest between 1670-1710, when temperatures fell to averages several degrees below current values, a cooling which coincided with a lessening of sunspot activity (the so-called 'Maunder Minimum') and reduced solar output. However, these were not years of unrelieved cold. Cool spells alternated with phases of milder weather. But these took place against a background of generally low temperatures, which would probably have continued to the present, had it not been for human activity.

Rainfall increased. There were large-scale advances in glaciers and ice sheets, and massive increases in the extent and range of sea ice. The Viking settlement in Greenland collapsed in around 1540 – which in itself was a miracle of endurance, as no shipping had been able to visit the place for nearly eighty years. The Viking settlement in Iceland also came under huge pressure. Its population withered away. Farming became a grim struggle for survival, and fewer supplies were able to reach the settlement because of icebergs and storms.

By the early eighteenth century the sea ice ranged so far south that on a number of occasions between 1690 and 1728, Eskimos, and in one instance a polar bear, made it all the way to Scotland. Rivers and coastal inlets across north-western Europe froze regularly – more than once a decade in some cases. During the winter of 1683-84, slabs of ice three miles in extent appeared in the English Channel, and the Thames froze at least twenty times during these years. Even the Mediterranean Sea around Marseille froze in 1595 and 1638, indicating that the big chill was not just a north European phenomenon.

There were more landslides and avalanches. Coastal flooding became frequent, and occurred on an unprecedented scale. Blown sand became a problem, particularly in the 1600s, indicating the destabilisation of the coastline as a result of more frequent and severer storms. The most astonishing example of this was the loss of the Culbin estate just outside Inverness in Scotland. Over a period of ten years, it was dramatically covered in sand to a depth of several metres, most of the overlay occurring during a single storm in 1695.[8]

Between 1660-1730, Europe's population fell, and life expectancy declined by as much as ten years. Farming, even more than usually, became a struggle to subsist, and livestock, wild animals and birds died in vast numbers due to cold and malnutrition. Widespread crop failures led to regular famines, particularly in northern Europe. Upland areas were again abandoned. Wine production ceased in parts of northern France because of the cold, and in the winter of 1708-09, Provence lost all its orange trees.

In Ireland, evidence of climatic severity abounds. In 1517 and again in 1683, the River Lee froze for weeks at a time,[9] and in 1715 there was a great fall of snow, which either lay, or continued to fall, for two months.[10] Matters reached their worst between December 1739 and September 1741, when the hardships inflicted by the Little Ice Age were horribly compounded by what seems to have been fallout from a volcanic eruption.

The 'Great Frost' of 1739-40 caused the Liffey, the Lagan, and Lough Neagh to freeze over. It produced the spectacle of banquets (and even a hurling match being held on the ice-bound rivers Shannon and Boyne). The only outdoor temperature reading taken during these seven weeks of aggressively cold weather showed some 32°F of frost.[11]

The rest of the year was also unusually cold. Manley's Central England Temperature Record shows 1740 to have been the coldest

During the Great Frost of 1739-40 rivers froze and mills ground to a halt. (Geoffrey Fulton)

year between 1659 and the present, with a mean annual temperature of just 6.8°C. This value is quite shockingly low. To put it in context, the lowest twentieth century mean annual temperature was 8.5°C (1963). The 2007 mean was 10.5°C, nearly 4°C higher.

This Siberian winter was followed by a prolonged drought, famine, typhus, dysentery and, in towns, the threat of civil unrest. Between 310,000-480,000 Irish people died during the crisis, out of a population of 2.4 million, a mortality rate greater than that experienced during the Great Famine of 1845-51. Writing in April 1741, as disease scythed through the malnourished population, the Rev. Philip Skelton from County Monaghan stated that:

> the dead have been eaten in the fields by dogs for want of people to bury them. Whole thousands in the barony have perished, some of hunger and others of disorders occasioned by unnatural, unwholesome and putrid diets.[12]

Numerous large storms are also recorded during the period. The most massive and most terrifying occurred on 'the Night of the Big Wind', the night of 6-7th January 1839. This storm, the worst in the Irish metrological record, holds a unique place in the folklore of the country, where it is represented as an event of almost mythical proportions.

Originating in a depression of around 920 millibars, which generated hurricane force winds, the storm left Dublin resembling 'a sacked city' and Belfast looking as though it 'had been reduced by artillery', five of its tallest mill chimneys having disintegrated during the night. Every parish in the country was affected, many severely. Over two hundred died, over three million trees fell, and incalculable damage was inflicted on property in what was probably the most ferocious storm to have hit Ireland in five hundred years.[13] How long that record will stand remains to be seen.

Records falling like ninepins: 1850 to the present

Between 1850 and 1920 temperatures yo-yo'd, as the Little Ice Age, as it were, wrestled with global warming. Would the descent into cold continue, or would global warming take charge? In the 1920s we got our answer. Temperatures took off, rising steadily up to circa 1945, which, in spite of the recent climatic epiphanies, remains the hottest year on record in Ireland.

This rise brought us out of the Little Ice Age, establishing new temperature planes in the process. But just when we thought we had got a clear fix on the new direction, we were wrong-footed again. The Little Ice Age kicked back. Between 1950 and the end of the 1970s or early 1980s, temperatures fell, prompting fears of a return not just to colder conditions but perhaps to a fully-fledged ice age. In the early 1980s, however, temperatures began to rise again until – against the background of rising sea-levels, increasing rainfall, and changing storm tracks – new record global temperatures were being recorded on an almost yearly basis. These four key ingredients of climate – temperature, sea-level, rainfall and wind – will be discussed in detail in the next four chapters.

4

Hotting up – temperature changes

The most worrying aspect of climate change is temperature rise, which has been marked from 1980 onward. Prior to that date, climatologists were concerned not with global warming but global cooling, and the imminent arrival of a new ice age.[1] And had there been no industrial revolution, this might well have been the case. But we have averted this catastrophe, and the fears of the 1970s read very ironically now. A pessimist might say we have jumped from the frying pan into the fire. An optimist would say that we are currently wrestling with the problems of success.

Global temperatures

The world's longest instrumental temperature record is Manley's Central England series, started by clergy with an interest in natural history in 1659. Its message is clear. All the hottest years in this venerable record have occurred within the last two decades. All the

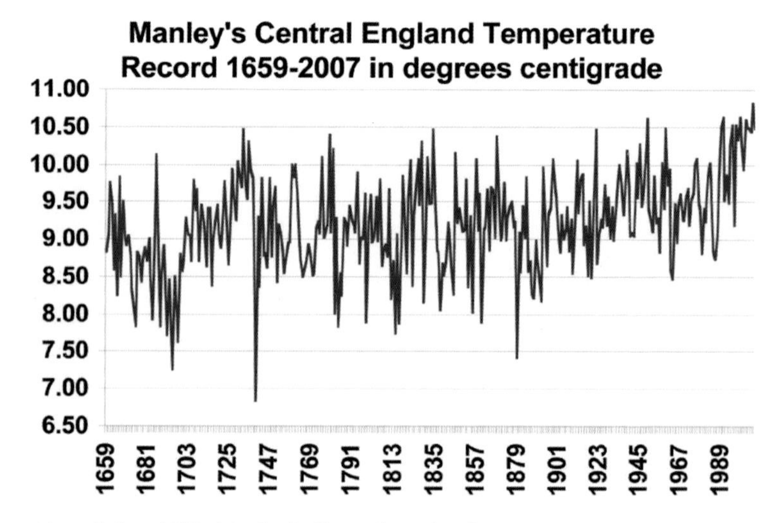

Founded in 1659, Manley's Central England Temperature Record is the oldest in the world. It shows rising mean temperatures.

coldest years, with the exception of 1740, 1887 and 1888, are in the first half of the record.[2] No year after 1888 has had a mean temperature of below 8°C. No year after 1986 has had one below 9°C, and since 2001 no year has had a mean temperature of below 10°C. A single degree may not sound like very much, but it is a very significant difference in this context. Central England is getting noticeably warmer.

The global trend is broadly consistent with this picture. Global temperatures have been rising, but not in a steady or statistically elegant manner. When annual global temperature anomalies are examined, they show that between 1850-1920, the earth's temperature fluctuated without showing any rising or falling trend. 1920-45 saw a period of warming. Between 1945-80 temperatures fell, but to levels that were much higher than those experienced fifty years before.

From 1980 onwards, we see the same sharp rise that was witnessed in the Central England series.[3] Since 1980 the earth has been warming at a rate of 0.2°C per decade. When it made its Third Assessment of temperature change in 2001, the Intergovernmental Panel on Climate Change (IPCC) found that, over the previous hundred years, the mean temperature had risen by 0.6°C.[4] When it published its Fourth Assessment in 2007, this figure was revised upwards to 0.74°C, a steep jump. Not only is the earth getting warmer, the rate of warming is increasing.

This rise has not been uniform. There have been significant regional variations. The 1901-2005 distribution data shows that the highest warming took place in the Alps, the western Mediterranean, parts of northern China and Mongolia, northwestern USA and Canada, and Argentina. Over the last twenty-five years the front runners have been Northern Canada, Greenland and Scandinavia, along with parts of Pakistan, China and South Africa.[5] The disproportionate warming of Greenland is of particular concern. Its huge ice sheet stores ten per cent of the world's fresh water. Were this to melt to any significant degree, none of the centres of the world's great port cities could be considered to be entirely safe.

What does the future hold? Is it as bleak as some environmentalists are predicting? The facts are these. Warming of 0.2°C per decade is projected for the next two decades. This is double the rate that was projected had greenhouse gas emissions been frozen at 2000 levels. Half of this will occur even if there are no further increases in greenhouse gas emissions; for the oceans are major reservoirs of heat and they will continue to release this heat into the atmosphere.[6]

Predicting future temperatures is very much a guessing game. However, sophisticated computer models, which factor in a wide range of economic, political, population and technological

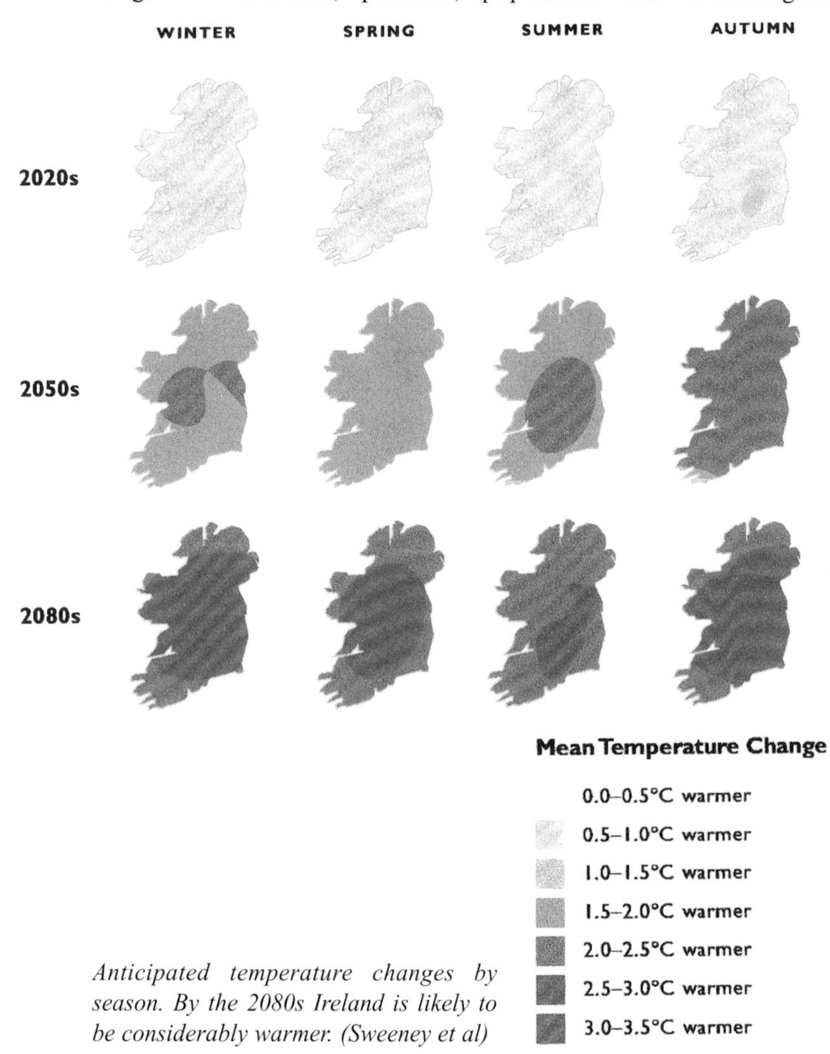

Anticipated temperature changes by season. By the 2080s Ireland is likely to be considerably warmer. (Sweeney et al)

contingencies, have been developed to, as it were, take the guesswork out of guessing. The most authoritative of these projections are set out in the Global Circulation Models, produced by climate research centres around the world.

These predict changes in mean global temperature of 1.8-4°C over the period to 2100, depending on the emissions scenario applied. This may seem slight, but it represents a significant increase in global temperature, and the increase will not be distributed evenly across the world. Some regions will warm by more than 4°C, others will warm more slowly. By 2025, temperature will rise by between 0.1ºC and 0.5ºC depending upon our success in controlling greenhouse gas emissions.[7]

The best composite, single-figure estimate deriving from this range is that global temperature will rise by around 3°C by 2100. This is not reassuring. Particularly not when we recall that last Ice Age was the result of a dip in global temperatures of just 4-5°C. 3°C may not seem like much, but it represents a massive and potentially overwhelming change.

Within the scientific community, there is general agreement that a 2°C rise represents a critical threshold, which we will exceed at our peril, the fear being that such a sharp rise within such a short time, will be more than we can adapt to. But it is about more than our adaptability. A rise of over 2°C will hugely alter the ecological base upon which the human superstructure rests. So 2°C has become a kind of *ne plus ultra*. Our ability, or inability to contain the rise to within 2°C may determine whether or not, in the very worst case scenario, human civilisation as we know it will survive.

Ireland – a climate change hot-house

Ireland's temperature has risen by about 0.7°C over the last century, with higher increases in the last few decades, a rise that is in line with the global picture. Again, this increase has not been uniform over time. From 1890-1920, Ireland's temperature was relatively stable. Then came a period of warming which culminated in 1945, the hottest year in the Irish meteorological record. After 1945, temperatures fell, then remained stable at a lower level, but a level that was higher than that of the 1890-1920 plateau. From 1985 onwards there has been a steady warming, but at a level that is *twice as fast* as the global norm.[8]

2007 was the warmest year on record at Malin Head, Valentia, Birr, Belmullet, Rosslare and Kilkenny. Eleven of the fifteen warmest years in the last century have occurred since 1990. Over the last hundred years, 2006 was the second warmest year, 1945 being slightly warmer, and the last ten years have been our warmest decade.[9] Something pretty remarkable is clearly underway. But what? The Community Climate Change Consortium for Southern Ireland Project was set up in 2003 to find out.[10] Its results, so far, have confirmed IPCC findings. Ireland is a climate change hot-house.

The table below shows projected Irish temperature increases. Rises of 0.8°C are expected by 2020, 1.5°C by 2050 and 2.3°C by 2080, meaning that we will sail through the 2°C point-of-no-return within the lifetimes of our children. All seasons show rises, but the rises are varied.[11] The biggest gainer will be autumn, followed by summer. The lowest risers are spring, then winter. These will be more than subtle changes. Irish weather will not be as it was.

Projected seasonal temperature increases
(in °C, after Fealy & Sweeney).

Season	**2020**	**2050**	**2080**
Spring	0.7	1.4	2.1
Summer	0.8	1.4	2.0
Autumn	0.7	1.5	2.4
Winter	1.0	1.8	2.7
Annual	**0.8**	**1.5**	**2.3**

Many of these changes will be – at first sight – benign. We can look forward to more Indian summers and milder winters. We may expect a greater number of hot days, more frequent and longer heat waves, fewer frosty days and fewer cold nights. The timing and extent of the warm up will vary. The warming of the interior and the south-east will slowly spread northwards, westwards and coastwards as the century progresses, with projected autumnal increases of up to 3.5°C throughout most of the eastern two-thirds of Ireland by 2080.[12]

Implications

These rises will touch nearly every aspect of life. It has been suggested, as a rule of thumb, that a one degree temperature rise equates to a 160 kilometre (hundred mile) southerly movement in latitude. Ireland will stay still, but by the end of the century, it will feel as though it has moved three hundred or more miles south.

That would put Belfast on the same latitude as Plymouth, leave Dublin enjoying the latitudinal advantages of Paris, and put Cork somewhere on the Loire. It doesn't sound so awful, and it won't be. Not for us. But you can get too much of a good thing. When we consider Madrid transposed to the latitude of Tangier and Rome transplanted to Tunis, we begin to see the potential downside, and the manifold implications for every aspect of life, from dress and fashion, to drinks consumption in summer and the incidence of skin cancer.

On the positive side, as winters will be shorter, with fewer really cold days, we will start heating our houses later and stop heating them earlier in the year. The great outdoors will be more welcoming. It will be possible to enjoy outdoor activities such as tennis and surfing for a longer part of the year. It may change the way some sports are played. In rugby, for example, we may see teams playing more of a running game, along the lines of that played by the southern hemisphere sides.

There will be a longer growing season, especially for grass, a trend that has been underway since the 1980s. Thc housing of livestock will be less crucial in winter, and may end entirely as the century progresses. Our flora and fauna are already changing. Warmth-loving

During the 2003 heat wave, UK temperatures reached 100 degrees Fahrenheit. (Emma Bayliss)

species are arriving. Cold-loving species are being edged out. In our seas, for example, gadid species like haddock and cod, which cannot survive in temperatures of over 17°C, are already retreating northward. Birds, being so mobile, are particularly responsive to temperature changes, and the environmental changes that they bring. Swifts, the last summer migrants to arrive and the first to depart, are staying longer, and as conditions become more benign, may even stay for good, as may other summer migrants.

Not every change will be welcome. The reduction in winter heating will be offset by the increased use of air conditioning, particularly in summer. Fans and cooling systems may become standard in homes, cars and offices. Indeed, in a couple of generations, we may crave shade in the way we now crave sun, and our homes may be as well supplied with fans as they are with radiators today.

Higher temperatures, especially at night, may cause rising numbers of fatalities. The 2003 European heat wave is a perfect example of what lies in store. Though no fatalities were recorded in Ireland, thanks in part to the moderating effect of the sea, over 2,000 died in Britain, and nearly 70,000 people died in twelve European countries as a result of this event, most of them elderly folk living in the continental interior, where day time temperatures rose above 100°F. 100°F temperatures were also seen in the UK for the first time since records began, along with night time temperatures that allowed people little chance to recover from daytime heat stress.[13]

There will be additional pressures on cold-loving flora and fauna including Arctic-Alpine plants such as Alpine Saw-Worth, Alpine Bistort, and Mountain Sorrel, and birds such as the Arctic Tern. Some of these species will become rare and others will become extinct here. Change is coming. There is no doubt about this. Only its scale and rate are in question.

5

'Every city a Venice' – sea-level changes

There is a clear link between rising temperatures and sea-levels. But not in the way most people think. For the widely-publicised ice-melt and retreat of the world's glaciers has not significantly affected sea-level. Getting on for 90% of sea-level rise, thus far, has been due to the thermal expansion of water. As water heats, it takes up more space, and the water in the world's oceans has been getting warmer – and will warm further. The fear is that this may usher in a new and altogether more formidable problem, the disintegration of the Greenland and Antarctic ice-sheets. Should this occur, we can expect sea-level rise on a catastrophic and unprecedented scale.

Global sea-levels

Over the last century, global sea-level has risen by between 10-20cm. When the nine best tide gauges in the world are averaged, they show a rise of just over 16cm between 1904-2003, or 1.6mm per annum.[1] The rate of rise is increasing. Between 1960-2003 the rate was 1.8mm annually. Between 1993-2003, it was a whopping 3.1mm per year. This may not seem like much in itself and is not unprecedented, but in terms of what had gone before, it represented a staggering acceleration, with potentially ominous implications for coastal regions. Furthermore, there is almost no glacial meltwater in this figure. The rise has been almost entirely due to the thermal expansion of seawater.

So what of ice melt? Why has it not counted? As we know, the world's glaciers are in retreat. Alaska's massive McCarthy Fjord Glacier, for example, an Amazon among glaciers, has retreated over sixteen kilometres (ten miles) in the last century, a rate of loss that is not untypical.[2] Fortunately, there has as yet been no significant loss of ice in the Greenland and Antarctic ice sheets. They remain intact. However, since 1995 there has been some shrinking around the edges, and an increase in seasonal melting.

Peripheral melting is an increasing concern. Two major ice shelves on the Antarctic Peninsula have collapsed and melted within the last six years. The Wilkes ice shelf started to collapse in 2008. Larssen B

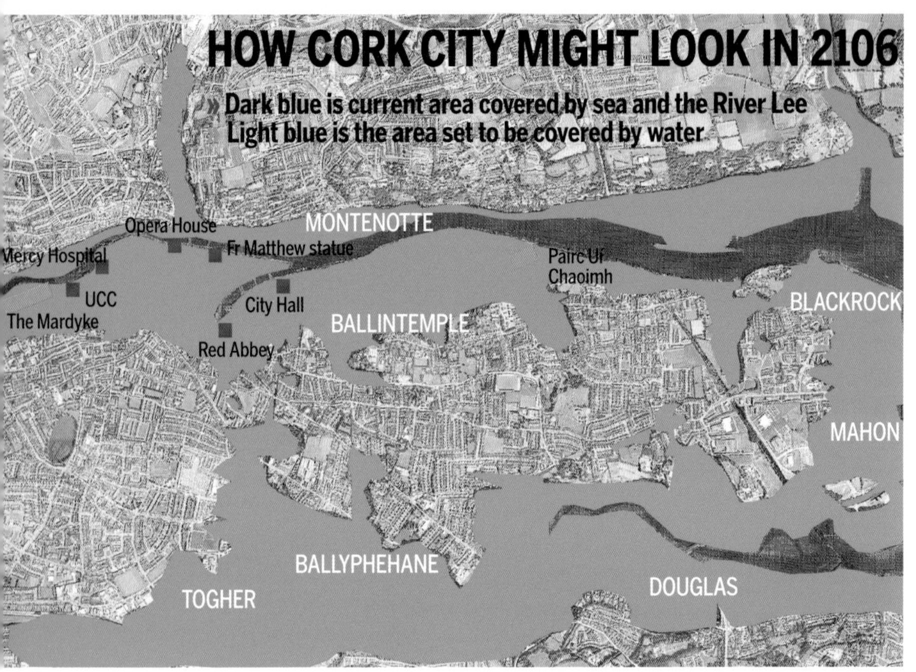

Cork after an eleven metre rise in sea-level. (Mapflow / Sunday Tribune)

collapsed over just thirty-five days in 2002. It covered over three thousand square kilometres (1,200 square miles), and consisted of five hundred billion tonnes of ice – the figures are staggering.[3] Ice shelves are huge blocks of frozen ice which float on water but are attached to the coastline, buffers, if you like around the land-based sheets. Fortunately, because these ice shelves floated on seawater, their melting did not raise sea-levels, an effect known as the Archimedes Principle.

In defiance of common sense, which tells us that oceans are level, and that sea rise should therefore be even across the globe, sea-level change is highly variable. The rate of change also varies significantly, with some parts of the world experiencing rates that are several times the mean global rise, while in other places, bizarrely, sea-level is falling. This is because the land itself is rising and falling at different points across the earth's surface, and because sea level is affected by ocean currents and the earth's rotation. We have also seen more storm surges, and what by historic standards are almost freakishly high tides. The incidence of these differs from place to place as well, according to variations in sea level and regional climate.[4] So the seas, then, are not 'flat' and settled, they are constantly changing.

By 2100, global sea-level is expected to rise by between 18 and 59cm, with a best guess estimate of 44cm (17 inches), two to three times the twentieth century increase. The way in which the rise is generated is not expected to change. The bulk will still come from the thermal expansion of sea water, topped up by melting ice.

These values assume no ice melt from the world's major ice-sheets. Were this to occur, we would be into completely new territory. Were the Greenland ice-sheet to melt, global sea-level would rise by around six metres, inducing a flood of biblical proportions.[5]

Without major sea defences, every city, in its turn, would become a Venice. Belfast, Cork, Dublin, London and New York would be hollowed out. They would lose their ports, their historic cores, their financial and commercial heartlands, their communication centres – all the things that make cities cities would disappear under water, leaving only a collection of anonymous suburban shells. Some cities in developing countries may have to be abandoned, as escalating costs made protection unaffordable.[6]

Until recently, climatologists had confidently ruled out this kind of scenario. Meltdown was impossible. Even the doom-mongers did not go further than the suggestion that the ice-sheet could break down over a period of two thousand or more years. But this confidence has recently taken a knock. Satellite monitoring of the Greenland ice-sheet has shown melting occurring in places where it was previously unknown, and it is now being suggested that significant melting could occur over as short a period as several centuries.

In Antarctica there are two ice-sheets to consider. The West Antarctic ice-sheet is the most vulnerable to melting. Were it to go, it would raise sea-levels by another 5-6 metres. The good news, if we can call it that, is that this ice sheet is slightly more resistant to melting than Greenland's, and were it to go, it would go more slowly.

Its big brother, the main or East Antarctic ice-sheet is an even more stable and intractable entity. This is just as well. Were it to melt, global sea-level would rise by around seventy metres, completely altering the world as we know it.[7] The consequences of this are too awful to contemplate. Even if only a tiny portion of any of these three ice-sheets were to melt by 2100, then sea-level would rise by beyond the predicted 44cm, causing potentially huge global problems.

Irish sea-levels: the north-south divide

Assessing Irish sea-level rise is a far from straightforward matter. This is because Ireland itself, as it were, does not lie still in the water. The north is rising out of the waves, and the south is sinking into them, while the island's middle reaches exhibit an almost stable 'relative sea-level'.[8]

These 'vertical' changes are small, and are a legacy of the last Ice Age.[9] A heavy blanket of ice then covered the northern third of the island, depressing those parts of the earth's crust directly beneath it, for the earth's surface is pliable, when pushed very hard. Upon the melting of the ice, like a plastic Coke bottle that someone had accidentally tramped on, the crust started to rise back towards its original position, a process that is continuing today. To the south of the ice blanket there was less crustal depression, and in the southern third of the island there was almost none at all. However, because sea-level rise is accelerating, it is possible that this historic trend could reverse, allowing the north of Ireland to begin to show a relative sea-level rise.

This may already be happening. Contrary to expectations, because northern seas are supposed to be falling, tide gauges installed at Bangor and Portrush ten years ago have shown a rather baffling sea-level rise. The Belfast tide gauge, however, using slightly older data, shows a sea-level fall of 0.25mm per year.[10] This tallies with expectations, as do the readings for Malin Head, which show a fall of 0.58mm per year.[11] The gauge at Dublin shows a rise of 0.23mm per year.[12] Again, this is more or less in line with expectations.[13]

These differences reflect complex adjustments in land and sea levels, and, just as importantly, the embarrassing poverty of the Irish data. Our understanding is based on no more than scraps of information. Although tide gauges have been in use in Irish harbours since the nineteenth century, there has been little or no analysis of their findings.

The lack of good Irish sea-level data is astonishing in this information-saturated age. However, attempts are being made to address the problem. An Irish National Tide Gauge Network has been set up, and in time this will produce usable data. The UK National Tide Gauge Network, which covers Northern Ireland, is grappling with similar issues. Ireland's historic tide gauge records, alas, are bedevilled by difficulties, including movement of the gauges, gaps in the records, haphazard and apparently slipshod recording – small wonder, perhaps, that these dusty volumes have not been published or digitised, or that we know so little!

What does the future hold? There are no bespoke sea-level predictions for Ireland, which leaves researchers in the invidious position of having to work with the anticipated 2100 global rise of around 44cm and apply this figure to Ireland, taking account of local land uplift and subsidence rates. For the reasons outlined above, the southern third of the island is likely to be worst hit, while the northern third, still vaulting free, is likely to weather the rise best. But it will not be immune, and even the reduced rise it may experience is capable of inflicting widespread damage. The Bangor and Portrush readings may be a straw in the wind. Though these records are too short to offer any certainty, they may show that sea-level rise has already overtaken land uplift in the previously 'safe' north-east.

An island fortress? – the implications of sea-level rise

As an island, Ireland is of its nature vulnerable to sea-level rises. Urban Ireland is particularly exposed, as most of the island's major towns and cities, including Belfast, Dublin, Cork, Derry, Waterford, Limerick and Galway, are by or within reach of the sea. And the sea is going to become an increasingly unpredictable neighbour.

Were the worst-case inundation scenarios to be realised, the results would be disastrous, as all of our coastal cities, bar Derry, are relatively low-lying. All bar Derry have recently experienced serious flooding, in some cases for the first time in generations.

In addition, Ireland has a 'soft' east coast, much of which is made up of unconsolidated glacial sediment (sand, gravel, clay, and loose rock), which is highly vulnerable to erosion. Only the rocky cliffs on the north-east, north-west, west and south-west coasts are anything approaching immune, shedding less than a metre every thousand years. But erosion rates increase with a rise in sea-level, so even these may become susceptible to more aggressive patterns of erosion.

As the sea rises, coastal and estuarine flooding will occur more often, and the areas affected by flooding will become larger, as the floods extend further and further inland. Water tables will rise, making all low-lying inland areas more vulnerable to flooding. Coastal water-sources may become contaminated with salt water, with serious consequences for the communities that depend on them.

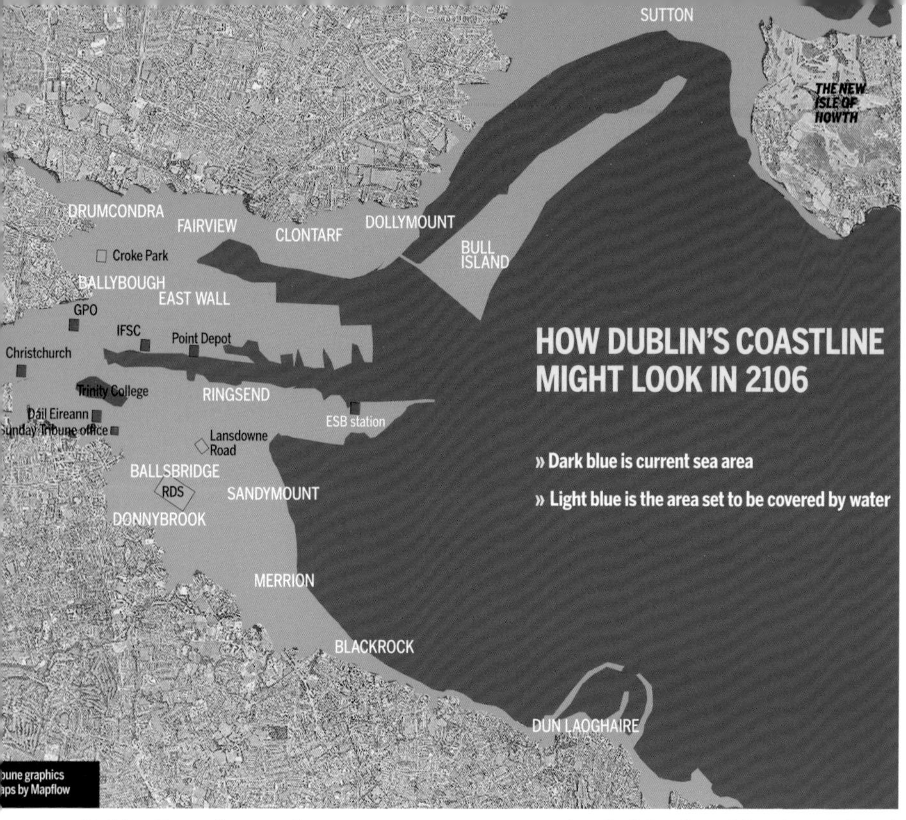

Dublin Bay after an eleven metre increase in sea-level. (Mapflow / Sunday Tribune)

These difficulties will be compounded by higher waves and storm surges, and shorter intervals between the last flood and the next, which in the next century could render life in small and repeatedly battered coastal communities potentially untenable.

Some idea of what a 44cm sea-level rise will mean can be had by examining the blip which produced Dublin's great flood of 2002. Examination of the high water levels for the period 1920-2002 shows that over this time only one flood was recorded. However, when the same data is superimposed upon a half metre sea-level rise, the result is not one flood over the eighty-two year time frame, but thirteen, a sobering indication of what the city can look forward to, unless we invest in the twenty-first century equivalent of medieval walling, and fortify the city against the sea.[14]

The coastline will also change. Rates of erosion will intensify, particularly on coasts made of unconsolidated material, which lie predominantly on the east. A study carried out in 1990 anticipated substantial rates of recession.[15] It showed that, given a 30cm sea-

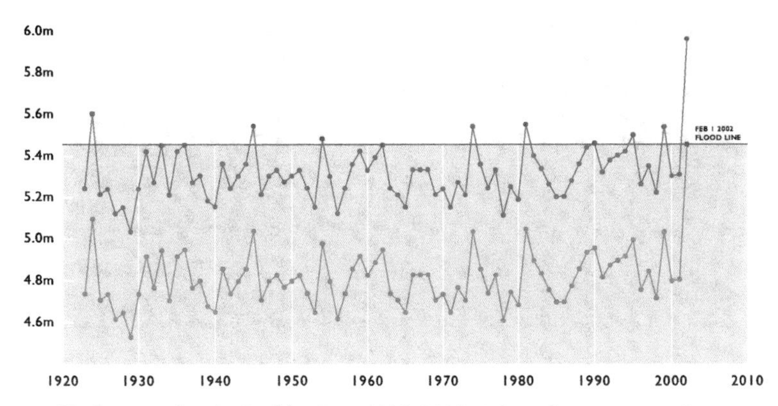

High water levels, Dublin Bay 1920-2002. When the anticipated 44cm sea-level rise is added, one flood turns into thirteen. (Sweeney et al)

level rise, which now seems very tame, average coastal retreat as high as fifteen metres could be expected by 2030 on the east coast. The sheltered north-east and the south-east would also have recession rates in excess of ten metres. Counter-intuitively, the exposed west, southwest and north coasts can expect recession rates of just two to three and a half metres.

However, it should be noted that these are averages only. Recession rates in vulnerable locations such as parts of County Wexford, the county with the biggest erosion problem, could be five to ten times these figures, with 'harder' locations being stranded offshore, and turning into islands. What does the future hold? In a thousand years' time, will our descendants see parts of the east coast of Ireland as we now see the coast of Norfolk? Off Norfolk's crumbling coastline lies a landscape of the imagination, a place of lost villages where, according to legend, the sea breeze carries the whispered peeling of submerged church bells.

It is astonishing to think that familiar Irish seaside villages could one day be viewed in the same way. But it is not impossible. The medieval town of Bannow, which used to return representatives to the old Irish parliament, is now under sand. The sea swallows about 750 acres of Ireland each year and the rate is accelerating.[16] Our coasts are already hard pressed by the forces of erosion. These pressures will intensify. Detailed management plans will be necessary if our beaches and sand dunes are to be maintained and enjoyed. But it will not always be possible to preserve well-loved strands, at least not in their current form.

Many beaches will be remodelled. They will narrow. Their character will change. Some will disappear entirely as their sediment is washed offshore. Barrow beach in Sutton, County Dublin, and Kilpatrick and Clones beaches in Castletown, County Wexford are already badly blighted and could be entirely lost very soon. The sand dune coasts of Waterford and South Down, and localities such as Magilligan Point are experiencing increased erosion. How ironic it would be to have warmer seas and warmer summers, but no strands on which to enjoy them! Fortunately, we are unlikely to be left without beaches. Many beaches will retreat inland, and new ones will form as deposited sand settles elsewhere, or covers areas that presently lie inshore.

The way we manage our 'softer' coasts may also have to change. Beaches and dunes serve as erosion buffers and it is important that we help them continue to do so. Activities which disrupt them, such as the removal of sand and gravel for building, and the use of vehicles on sand dunes (which can unsettle the sand, encouraging massive blow-outs), may have to cease or be curtailed.

But above all, we will need to spend money. A lot of money. Northern Ireland currently spends five million pounds annually on coastal defence. The Republic spends twenty-five million euro. This is nothing compared to what is going to be needed. In the wake of

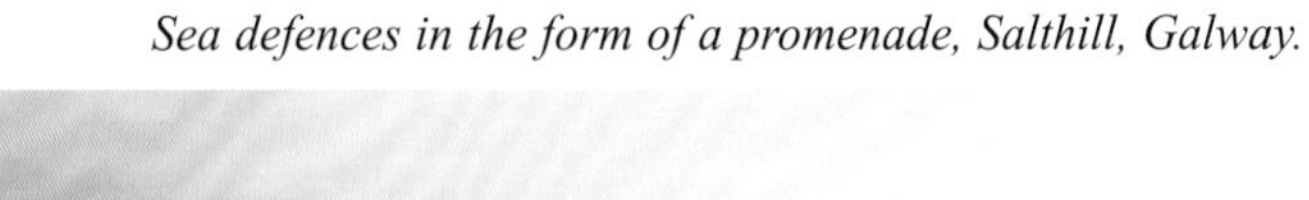

Sea defences in the form of a promenade, Salthill, Galway.

Hurricane Katrina, the US government initiated a fifteen billion dollar flood defence programme. But this was closing the stable door after the horse had bolted. Leadership is not about responding boldly to disasters. The trick is to have flood defences in place *before* disaster strikes. Expenditure of the order of several billion euro will be required over the next two decades, if Ireland is to protect its coastal towns and cities and vital commercial and other infrastructure.

Hard questions must be asked. Is the vulnerable Shannon Estuary adequately protected? Was the Belfast Barrage built taking sea-level rise into account? Will the nineteenth century sea walls around many of our cities be equal to the twenty-first century attack? The coast is also a focus of investment in transport and power infrastructure. Are our railway lines, airport runways, and power stations high enough above projected sea level? The answer may in part be no, in which case serious consideration needs to be given to moving or protecting them.

Hard engineering solutions, however, have their own problems, and it is easy to solve a problem on one part of the coast by creating another elsewhere. Certainly we shouldn't throw taxpayers' money at defending agricultural land, or even golf courses. The Wexford County Development Plan bars building closer than a hundred metres to the coast, a wise precaution. Planning regulations across the country will need to be adapted to a situation in which every centimetre rise in sea-level can equate to a retreat of one metre of land.

Speaking at the launch of his 2004 report on future flood risks in England and Wales, Sir David King said that the UK should take the threat of climate change and flooding as seriously as it takes the threat from terrorism. This is a prophetic warning. At the moment however, flood defence lies some way behind dog fouling as a political priority. It needs to be raised up the political agenda. We are barely coping with existing flooding problems, never mind adequately preparing ourselves for the difficulties that lie in store.

But there is an irony here too. The relief such measures can offer is temporary. There is a futility about them. The hard truth is that neither government, north or south, will be able to protect the entirety of its coastline because of its length and the colossal costs involved. Not now, but at some point in the far future, hard choices will have to be made. At what point will government decide that such and such a village should be abandoned to its fate? A village that has perhaps had millions already spent on it, and lies in the constituency of a government minister?

Protection works can cost from several hundred thousand to several million euro or pounds per kilometre. These costs are likely to be prohibitive. It will not be possible, or desirable, to turn the island into a fortress. The cruel reality is that we will have no choice but to leave large tracts of agricultural and other relatively low value land to the mercy of the sea.

6

Forty shades of yellow – rainfall changes

Rainfall, in whatever form it takes, is crucial to life on earth. The amount that falls varies hugely from place to place, and climate change has added a new twist to the pattern. Irish rainfall is going to become markedly more seasonal and this will have major implications for the way we live our lives.

Global rainfall

Will a warmer Ireland be a drier Ireland? Unfortunately not. A warmer world is likely to be a wetter world, because increased temperature leads to increased evaporation, which produces more cloud and more downpours.

Warmer atmospheres also tend to hold more moisture. Global atmospheric water vapour levels and the frequency of heavy rainfall have both increased as a result of the recent rises in temperature.[1] Between 1901 and 2004, global rainfall rose by a figure of between 11-21cm, a significant increase in a global context.[2] This rise peaked in the late 1950s, after which came a short-lived decline. The 1970s saw another peak, followed by a decline into the early 1990s. Then rainfall started rising again.

This bounty is not evenly distributed. Regions in the mid to high latitudes – including Ireland, Britain and Europe – have been receiving above average increases. Tropical and sub-tropics areas, including sub-Saharan Africa and other drought-ravaged regions, have been getting less rain. This decline has produced longer and more savage droughts in many of these places.[3]

Between now and 2100, the strong twentieth century rate of increase is expected to continue, thanks to the ongoing heating of the oceans and the rain cycle, as described above. Regional differences will intensify. Europe's skies will be cloudier and more glowering in 2100. There will be more rain. Areas around the equator will also receive more rain, while sub-tropical regions will receive even less. If there is a rain-based horror story in the climate change narrative, it will lie here, in the struggle of increasingly rain-starved places like the Middle East and Australia to manage with less water. Changes in run-off will follow roughly the same pattern, and where there are increases in run-off, such as north-west Europe in winter, there is likely to be more flooding.

Irish rainfall – increased seasonality

Ireland is famous for its rain. It falls copiously, in the form of everything from soft mists to stair-rods, and it will come as no surprise to non-Irish speakers to learn that there are seventy-eight terms for rain and rainy weather in the Irish language.[4]

The trends in Irish rainfall are variable. Over the last hundred years, Malin Head in County Donegal has received steadily increasing amounts of rain, whereas Birr in the centre of Ireland gets much the same amount as it did a century ago.[5] This mirrors the general trend, in which the North and West have had more days with heavier rain, with no clear pattern of change elsewhere. A recent study found increases of the order of 15-20% along the west coast, with very much smaller increases elsewhere, and no change at all in Dublin's rainfall over the last forty years.[6]

By 2080, these trends will have produced dramatic disparities. Irish rainfall will acquire an emphatic seasonality. Winters will become very wet by today's standards, with 15% more rain falling. Sodden fields, brimming rivers, waterlogged golf courses and messy encounters at Lansdowne Road can be expected.[7] Were there to have been no rise in temperature, we might have expected huge fluffy snowfalls, tobogganing children, and cars abandoned in snowdrifts. As it is, snow will become an increasing rarity. Bets on a white Christmas will attract longer odds. Snowy yuletide scenes on

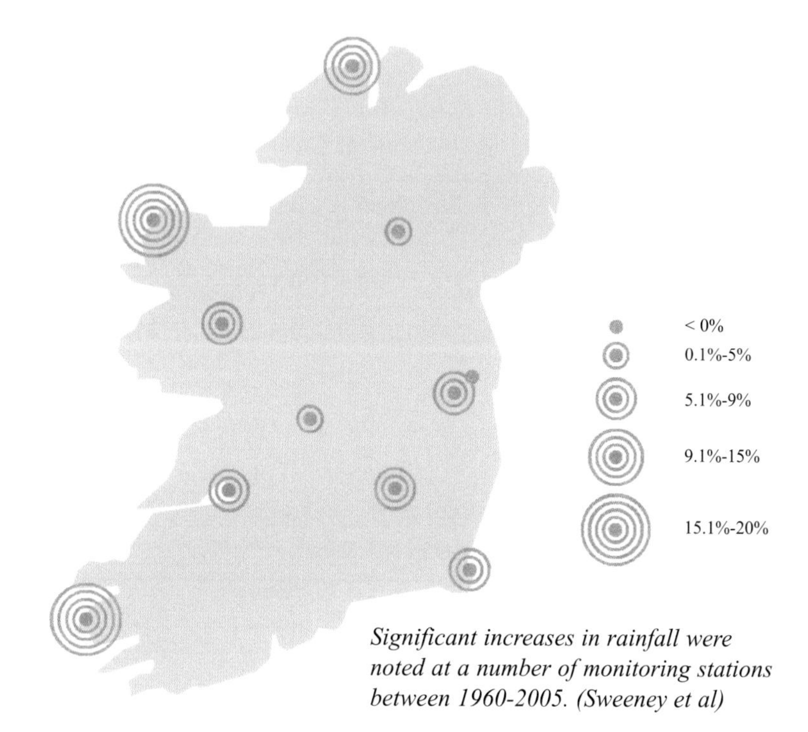

Significant increases in rainfall were noted at a number of monitoring stations between 1960-2005. (Sweeney et al)

Christmas cards will become objects of nostalgia and fond remembrance.

Spring, summer and autumn will become drier. Summer showers will become rarer, with about a fifth less summer rain falling by 2080. But we needn't pack away our umbrellas just yet. Analysis suggests that we can expect more downpours and freak deluges, particularly in the winter half of the year, which will also see more two-day long deluges, especially in western, south western and north western areas.[8]

Projected percentage changes in rainfall
(*relative to the 1961-90 mean, after Sweeney et al*).

Season	2020	2050	2080
Winter	3.0	12.4	15.6
Spring	-1.0	-7.2	-8.0
Summer	-3.2	-12.1	-19.0
Autumn	-1.7	-2.6	-7.1

Implications

This rainfall will inundate our streets and fields, causing mayhem, but ultimately, it will find its way into our rivers. A recent study of the Boyne has shown that dramatic deviations from the historic flow pattern can be expected in the future.[9] The river will become a torrent in the depths of winter, carrying a fifth more water than it does at present. A recent hydrological study of the Suir also noted an increase in the number and intensity of its overflows.[10] Some projections suggest that the once in a century flood may occur as often as once in a decade. All the indicators suggest that more flooding is inevitable and this must be planned for.

In summer, the reverse can be expected. By 2080, a large tract of the south-east, including counties Cork, Waterford, Wexford, Wicklow, Dublin, Louth, and parts of County Down, is expected to receive less than half its current summer rainfall. The Boyne, along

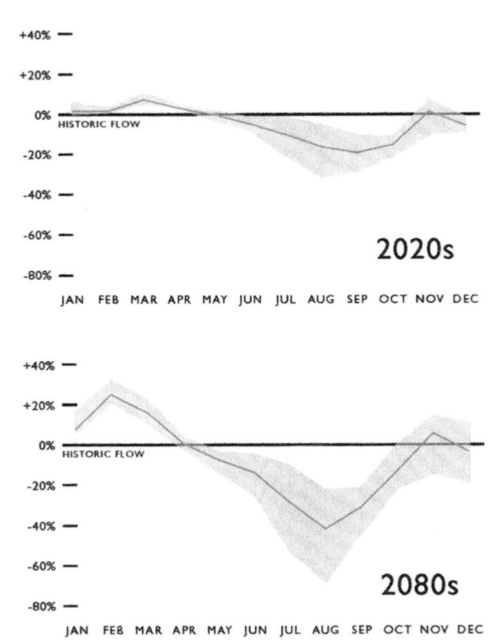

Dramatic changes in river flows can be expected. As these charts show, by the 2080s the River Boyne is likely to lose as much as 40% of its current summer flow.

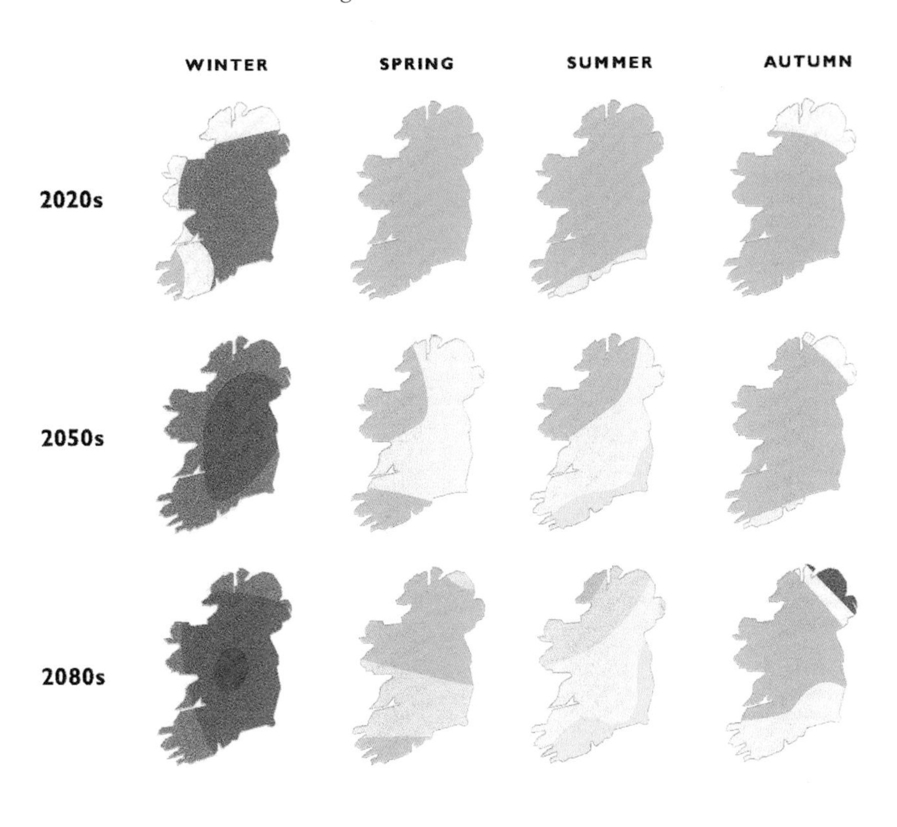

Precipitation (% Change)

- > 51% less rain
- 41% – 50% less rain
- 31% – 40% less rain
- 21% – 30% less rain
- 11% – 20% less rain
- 1% – 10% less rain
- 0% no change
- 1% – 10% more rain
- 11% – 20% more rain
- 21% – 30% more rain

Anticipated rainfall changes by season. Our summers will become noticeably drier and our winters very much wetter than at present. (Sweeney et al) with the island's other great rivers, will shrink to a shadow of its current self, exposing areas of riverbed that have never been exposed before. These changes will delight archaeologists but will pose practical problems for the authorities, not the least being how much effluent we can continue to dispose in them.

The decline in rainfall will also have major implications for water resource management. This will especially be the case in the relatively crowded east, and the problems could be acute in Dublin and Belfast and other eastern towns and cities, for the areas of greatest demand are going to receive the lowest rainfall. Demographic trends will exacerbate these difficulties. Around an additional 750,000 people are expected to live in the Greater Dublin area by 2020, and population increases running into hundreds of thousands are anticipated in the other east coast settlements. Water consumption per person is also rising, putting additional demands on supply.

Will our eastern cities have enough water? The signs are that Dublin in particular will need a huge new supply within the next decade, most likely from the Shannon, with all the local sensitivities and disruption that this will entail. Only Dublin's closeness to the Wicklow Mountains and Belfast's proximity to the Mournes, both of which receive copious rainfall, have prevented this issue from coming to the fore until now.

Water conservation, then, will become a priority. Wasting it will become increasingly unacceptable. In Northern Ireland, around 157 million litres of water are lost each day through leaking water pipes.[11] Twenty million litres are lost daily in Dublin. Public opinion will demand that we improve our leaky systems, not withstanding the huge cost of replacement.[12] During summertime water users – on farms, in commerce, and in the home – will find themselves competing for a scarce resource. Difficult decisions will have to be made, with some users and uses being prioritised over others. The political and social tensions that this will involve can be imagined.

The increased seasonality will have a profound impact on Irish agriculture, and the type of crops that are produced. Crops that need regular summer watering, such as potatoes, may give way to crops that require less, like maize. This could transform the Irish farming landscape. Metering may become more widespread. Farms with easy access to water will become highly prized. Livestock farmers will need to make special provision to ensure that their animals are adequately watered during the drier summer months.

The natural environment will also feel the impact of this schizophrenic weather. Winter gluts and summer dearths will test Ireland's entire ecology. In winter, landslips and bog bursts will become more common. In summer, many smaller streams and rivers, which at present just about manage to carry their marine life through the summer, will dry out or be reduced to strings of semi-stagnant pools, crowded with pond life, which are as likely to be refreshed by rainfall as any flow from upstream.

Peat bogs are also likely be stressed by longer periods of lower rainfall, as will the flora and fauna they support. Water-loving plants will find it increasingly difficult to survive the summer. As our forty shades of green give way to forty shades of yellow in late summer and early autumn, spontaneous gorse and forest fires will become more frequent. This will make demands on the third of our emergency services. We have already envisaged the police being called out to attend to disputes between neighbours over water, and the ambulance service being called out to revive heat stroke victims. As our rainfall becomes increasingly bi-polar, the fire service will also be kept on its toes.

7

Stormy weather – wind changes

Storms are the biggest natural hazard Ireland faces. Every year, we are hit by three or four big storms which cause significant damage and occasionally loss of life. This chapter looks at the forces that cause our storms, and how they are being affected by climate change.

Global wind and storms

Storms exert a hypnotic fascination. Our fear of them is almost primal. And well founded, for they are capable of bringing death and destruction on a massive scale. The meteorological inter-

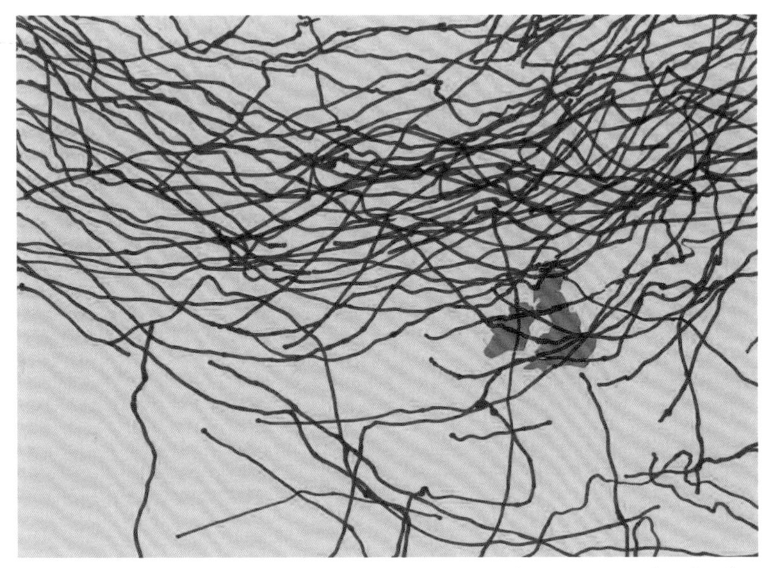

The 1988-89 storm track pattern, in relation to the Britain and Ireland.

relationships which trigger storms, hurricanes, and cyclones are, in their way, no less fascinating. And over the last few decades, global warming has 'tweeked' these complex inter-relationships to produce three notable changes in storm activity.

The most interesting change has been in what is called 'storm track location', or in plain English, where storms occur. Over the last few decades there have been fewer storms in the mid-latitudes, and more storms in the high latitudes. Detailed analysis of the data suggests that the winter North Atlantic storm track, our storm 'superhighway', has moved about 180 kilometres (110 miles) northwards since 1950.[1]

As most of Ireland lies in the mid-latitudes, this development has – broadly speaking – been good news. The storm index for Ireland, Britain, the North and Norwegian Seas for the period 1881-2007 confirms this view, showing a bumpy but almost continuous decline in storminess from 1881 to the present.[2] So, to put it into plain English, the wreckage-strewn storm superhighway is ceasing to be a through road, and becoming a by-pass. The storms are happening up north somewhere, and increasingly leaving us alone.

The second and rather less welcome change has been an increase in storm intensity. The world has seen a significant increase in the destructiveness of hurricanes since the mid-1970s. Storm intensity and storm duration have also been increasing over the North Pacific and North Atlantic. We are getting not only rougher but longer storms.

Thankfully, the third change is more cheering. Contrary to the general perception, the last few decades have seen a slight but definite *decrease* in the total number of storms.

These changes have not been confined to the mid and high latitudes. Tropical storms and hurricanes have also been changing in frequency and intensity, but in ways that are hard to make sense of, as their character is so variable. Their frequency and behavior is also influenced by the notorious El Niño, which regularly reverses wind currents in the Pacific, produces drought in Australasia, and increases hurricane activity on the western side of Central America. Its eastern counterpart, La Niña, generates hurricanes in the Gulf of Mexico.[3]

The most powerful hurricanes are those which fall into Category Four (wind speeds of 183-217 kilometres, or 115-135 miles per hour) and Category Five (speeds exceeding this). Disconcertingly, their number has increased by three-quarters since 1970. The year 2006 was relatively quiet, thanks to the El Niño. But 2007 saw two Category Five hurricanes.[4] And in 2008, a massive typhoon, the worst in Burma's history, claimed at least 133,000 lives and devastated a large part of the country.

North Atlantic hurricane numbers have also been above average in nine of the last eleven years, culminating in the record-breaking 2005 season. This included an astonishing *four* Category Five hurricanes, amongst them Hurricane Katrina, a storm the size of France, which scythed through New Orleans, causing 1,836 deaths and forty billion dollars worth of damage. The North Atlantic has not been this stormy in a long while. We know this because its storminess has long been carefully monitored, and typically found to 'tick over' at a rate of three Category Five hurricanes per decade.

It would be reassuring to think that this avalanche of larger and more ferocious storms has been a blip in the record and will pass. But there is no evidence that this is the case. The likelihood is that, paradoxically, as global warming makes the tropical troposphere more stable, tropical cyclones will become more severe, achieving flailing wind speeds and spewing out ever larger amounts of percussive, storm-driven rainfall. The havoc they wreak, the number of lives they claim, can only increase, it will not get smaller.

Irish wind and storms – less means more

There have been dramatic changes in Irish storminess over the last two hundred years or so. These have been carefully tracked by meteorologists in Armagh, Edinburgh, and Valentia, County Kerry.[5] The Armagh weather diary shows significant long-term variations in the storm record, with high numbers of storms recorded in the early 1800s, 1950s, 1960s and 1980s, followed by a dramatic fall-off to the present. The Armagh weather observation data shows an irregular but strongly cyclical storm pattern that is apparently influenced by the North Atlantic Oscillation (see below) amongst other variables.[6]

The Valentia hourly wind data more or less confirms this quasi-cyclical pattern (its peaks and troughs are even, very helpfully, in more or less the same place). It builds on the Armagh data by setting the pattern within a general fall in 'gale hours', a fall that becomes a tumble after the mid-1980s high,[7] which of course is in keeping with

Woodland laid low by the 1987 storm.

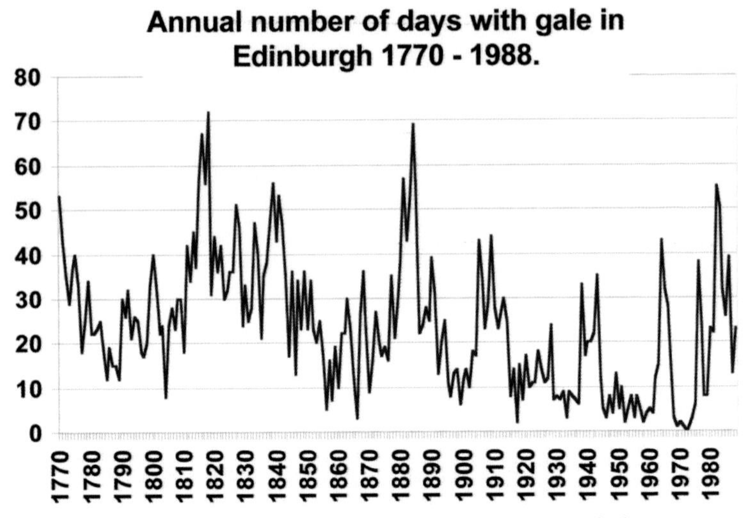

The three most turbulent periods on the record coincide with the eruptions of Tambora (1815), Krakatoa (1883) and El Chichon (1982).

the poleward movement of the North Atlantic storm track.

Volcanic eruptions have also exerted a powerful, occasional influence on Irish storminess by increasing atmospheric volatility. This is clear from the Edinburgh gale record. Its three most ferocious bouts of storminess immediately follow the eruptions of Tambora, Krakatoa, and El Chichon in 1982. Even though they occurred in far off lands, these catastrophes significantly increased the unpredictability of our weather.[8]

The worst storm on the Valentia record is the gale which struck the south-west corner of Ireland on the 26th-27th December 1951. This storm was remarkable for its longevity. Winds reached or exceeded gale force (Beaufort Force Ten) for an astonishing eighteen hours. The storm's sheer longevity and fetch generated some of the highest waves on record on the west coast of Ireland. Skellig Rock Lighthouse was hit by a wave estimated at over 36 metres (117 feet) high, as it knocked out the light and smashed numerous glass panels on the top of the lighthouse, terrifying the lighthouse keepers. In Kilkee, County Clare, a succession of waves, which arrived at five in the morning, washed over a row of one-storied houses on the sea front. The event lives on in local folklore, where it is recalled that seawater poured down people's chimneys, causing consternation.[9]

What lies ahead? The predictions for Ireland very much match those for mid and high latitudes around the world, with the predicted number of very intense superstorms doubling by the end of the century. On the positive side, for it is not all woe, it is anticipated that we will probably experience fewer storms, thanks to the poleward shift of the storm track, which, in this regard at least, may prove Ireland's salvation.[10]

These storms will be no more predictable than they are at present. We will have years with few, and years with higher numbers of storms, the latter coinciding with a strongly positive North Atlantic Oscillation, that is to say a higher pressure difference between the Azores low and the Iceland high, the bigger the difference, the greater the likelihood of storms. This phase of the cycle is the one we will come to dread, for it will mean more and bigger storms, accompanied by warmer and wetter conditions.[11]

When it comes to wind and storms, then, there are as many positives as negatives on the balance sheet. On the credit side, Ireland can expect to enjoy a generally more benign wind environment. The

Choppy seas. Gale force winds strike Kircubbin in January 2006. (Press Eye)

difficulty is that when it gets bad, it will be awful, with an increase in the types of storms that are capable of causing death and widespread destruction. In the 1940s, Ireland's level of forestation was a derisory 1.5%. Over the last sixty years, the figure has crept up to 9%. A really bad storm could knock that figure back by a couple of per cent in just one night.

These more exuberant storms will also maul the coastline, as storms of this sort of potency can cause significant erosion and flooding, both on the sea coast and in estuaries, as Cork and Waterford know all too well. If we need reminding of just how devastating major storms can be, we need only look to Burma. All that remains is to batten down the hatches in anticipation of the Night of the Big Wind, Mark Two.

8

The natural world

This chapter outlines some of the main changes that are taking place in the habitats, flora and fauna of Ireland, and examines what we might expect in the future.

Threatened habitats

Whilst a great deal of effort has been invested in modelling the impact of climate change on ecosystems elsewhere, relatively little work has been done on the consequences of global warming for Irish biodiversity. Compared to those of Britain and the continent, Irish ecosystems are neither complex nor diverse, which may lend them resilience in the face of climate change.[1] Conversely, Ireland's relative isolation may actually make its impoverished flora and fauna more vulnerable to infiltration by alien plants and animals.[2]

Although a lot more research is needed in this area, it is clear that changes are inevitable in key habitats such as salt marshes, turloughs, machair and coastal lagoons, as well as on mountains and peatlands. Some of these habitats are precious assets – European or even global treasures – which we can ill afford to lose.[3] In 2008, fifteen Irish habitats were identified as being under threat.[4] Most lie on the coast, or are based around lakes and wetlands. Ironically, the biggest threat to them comes not from climate change, but from more workaday human activities, such as farming, suburbanisation, and infrastructural development, to name but a few. However, climate change is increasingly muscling into the picture to further diminish their chances of survival.

An Ireland without bogs is a strange thing to contemplate. But increased decomposition of peat, resulting in the bog bursts we have seen in recent wet periods would appear to be likely as this great carbon store begins a slow disintegration. Increasingly seasonal rainfall will lead to marginal bogs drying up in summer, and eventually disappearing. By 2075, it is estimated that almost half of our peatlands will have gone. These changes in habitat will mean changes in plant, animal and bird species. Plants like Bog Cotton, which as their rich folklore testifies, were once bound up with human society, will be put at risk. Lakes, wet heath, fen lands and lowland

'An Ireland without bogs is a strange thing to contemplate...'
The Cotton Bog, Ards Peninsula, at dusk.

hay meadows will be severely pressured. Sand dunes and beaches are also vulnerable. Rising sea-levels and harsher storms will erase and alter many, imperilling coastal plant communities.

But it is not all gloom and doom. Most plants and trees are enjoying a longer growing season, a trend strikingly confirmed by satellite photography.[5] Grasses are doing particularly well. The grass-growing season is currently lengthening by five days every decade, and the hardier grasses will thrive under the forthcoming regime, provided they get enough water to see them through the summer.

Plants

Irish biodiversity is something of a revolving door. It gets pushed one way, then another. Species enter, species exit. Things never stand still. At present, Ireland is becoming more welcoming to warmth-loving plants, and increasingly intolerant of those that love cold. Plants that enjoyed its mild and temperate character are in headlong retreat northward. Some are already being elbowed out. Plants that once found Ireland too cold are showing an ever-livelier interest.

This is changing the ecological make-up of the island. Eleven native plants, most of which have long had a fairly tenuous grip on the landscape, have recently become either extinct, or extinct in the wild, and climate change has played a major part in their demise.[6]

The list of threatened species is much larger. 171 of our 850 or so native plants (around 20%) could disappear by 2080 as a result of climate change, a purge on a scale unknown since the last Ice Age.[7] The list includes curios such as Irish Lady's-Tresses, Weasel's Snout, Dwarf Spike-Rush, and the Corky-Fruited Water-Dropwort. Are any of these familiar? Would you recognise any? I'd be most impressed if you did. And their unfamiliarity tells its own story. Most of the listed plants live a precarious existence. Some are just about hanging on.

Many of these species have common characteristics, for the list is something of a roll-call of niche species, plants that are tied to restricted or vulnerable habitats. It also includes plants that are vulnerable to invasive rivals, plus those most likely to be affected by delicate changes in the ecological balance, such as a fall in the number of insects that they depend on to pollinate them.

Even so, their loss would be enormous. At a stroke, the Arctic-Alpine component of Irish flora would be decimated. Cold-loving

Facing an uncertain future. The Irish Lady's Mantle. (Emma Bayliss)

species, such as those native to the Burren and other limestone and upland areas, will become more restricted in their distribution or disappear. Prime amongst these are the hardy Alpine Meadow-Grass, the Mountain Avens, a ground-hugging shrub, the flowers of which make a vivid green dye, and Alpine Lady's Mantle, traditionally used for cleaning wounds and menstrual problems.

Some native species are already down to a handful of individual plants, living 'secretly' in locations known only to a small number of botanists. Climate change could cause their extinction in Ireland. In the case of the Killarney Fern, extinction here would mean extinction globally, as it is that rare thing, a uniquely Irish plant.

Many of these endangered plants are legally protected.[8] The seeds of others are held in the recently opened Global Seed Vault in Svalbad on the remote island of Spitsbergen in the Arctic Circle, which stores the seeds of threatened and other plants from around the world. The National Botanic Gardens in Glasnevin are also actively conserving rare and endangered Irish plant species. These botanical 'Noah's Arks' have perhaps never been more necessary, owing to the loss of some seedbanks due to war and political instability. The World Wildlife Fund has warned that over a hundred thousand plant

Colourful colonist. The ripening fruit of the Strawberry Tree.

species are under threat worldwide.[9]

Thankfully, the traffic is not all one way. Warmth-loving species like the Mediterranean Heather and the Hairy Saxifrage will thrive in the new Ireland. Another that may do well is the Strawberry Tree, a large, leathery-leaved bush, the fruit of which has been likened to a poor man's strawberries. These are currently restricted to the south-west corner of Ireland, and a few sheltered locations along the west coast, but are likely to spread northwards and inland as temperatures rise.

The seeds of warmth-loving species will be blown or carried here. In the past, conditions were too harsh for these to germinate, but this may not be the case in years to come. It is hard to guess what will arrive, but the potential for change is enormous. Our species pool is small. Thousands of species that are not found in Ireland live along the north-west fringe of Europe, many of which could arrive and spread rapidly as conditions improve.

Frost – or its absence – is in some ways the nub of the issue. Frosts, their number, their severity, but above all, their timing, will play a big part in determining the success or otherwise of these invasions and 'break outs'. For it is our night-time, not our day-time temperatures, that matter to most would-be colonisers. Cool nights and hard frosts have long held potential warmth-loving settlers at bay. But Jack Frost is no longer the terror he was. He is no longer the sentinel at the door. Frosts are becoming scarcer. Inland, where more air frosts form, the annual count of frosty days has fallen by ten to fifteen since 1940. Along the coast, where frosts are fewer, the fall-off is around five to ten days, the most dramatic drop being at Rosslare, which rarely has a frosty day, a change which opens the way to all sorts of invaders.

The Hottentot Fig, which is as yet known at only twelve Irish sites, all lying between Cork and Belfast Lough, is one such potentially dangerous invader. A single plant can grow to fifty metres across, and its effects are pernicious. It out-competes and smothers coastal flora, and as the climate warms this densely matting plant is likely to spread rapidly along the eastern and southern shores of Ireland, as it already has in southern England. The Invasive Species Ireland Project (a joint north-south venture) warns that other pestilential species are also waiting in the wings. These include the Giant Salvinia, which has been known to choke up canals and even reservoirs, and ranks high on the US government's Federal Noxious Weed list, the Tree of Heaven, the Empress Tree, the Fanworth, and the Black Locust Tree. All are already in Britain, and may well come

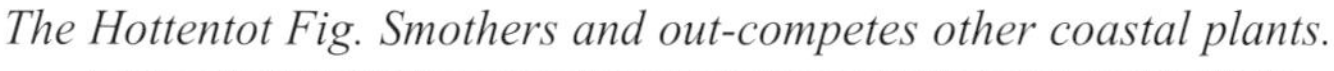

The Hottentot Fig. Smothers and out-competes other coastal plants.

here next. However, climate change will not give Mediterranean species *carte blanche*. Those that struggle with high rainfall and waterlogged soils will not flourish.

So much for the exotic invaders. What about the species that surround us, the plants we know and love? Most will remain, but their numbers may change. Many of our deciduous trees, including oak, ash and elm, will do well out of climate change, so long as they get the necessary water. Others such as beech will be hit hard by the drier summers, and will not do so well. Trees like birch and willow and many conifers, which enjoy our cool summers, may also fare poorly.[10]

Animals – *plus ca change*

A group of distinguished scientists recently attempted to assess the impact of climate change on the world's plant and animal populations. They inquired into the fortunes of some 28,800 species, so their survey was nothing if not comprehensive.[11] The results were startling. They found that at least 90% of environmental damage, including habitat loss, could be explained in terms of climate change. Every continent was affected, as was every class of animal. In 2006, the World Wildlife Fund issued an equally sobering message, calling climate change a significant threat to species worldwide.[12]

The message is one of turnover and change. However, we are not seeing this in Ireland. We have not experienced alterations in the variety or abundance of our animal life that can be attributed to climate. At least, not yet. This is because of the nature of our wildlife. Ireland is largely populated by 'generic' mammals, mainly cross-European species such as the Badger and the Pine Marten. These are not confined to niche habitats, which makes them better able to adapt to change, making it unlikely that climate change will greatly affect them. If anything, the warmer conditions may boost numbers, especially of rats and mice, which could become rampant.

There have been newcomers. Recent discoveries include populations of Muntjac Deer, which were introduced to be hunted, and the Greater White-toothed Shrew, found earlier this year. Its arrival is good news for Barn Owls, but bad news for its smaller native rival, the Pigmy Shrew. Neither of these introductions can be attributed to climate change.

Nor are any foreseen, at present. For once, the prognosis is refreshingly bland. Climate change will not be shaking up our mammal populations, at least not any time soon. This is because of their generalist requirements. Ireland's relative isolation, as an island off an island, adds a further barrier to species crossover. The threats are there, but they come from other quarters. In the short term at least, our mammals have much more to fear from old staples such as rising population, urbanisation, rural over-development and coniferous afforestation, than the impact of climate change.

Birds – the wings of change

Coal miners took canaries underground for a very good reason. Birds are extremely sensitive reflectors of differences in climate and habitat, and make excellent barometers of change. And they are just as likely to be affected by climate change as any other group of Irish animals. However, unlike most animals, birds can easily escape to more comfortable climes.

Birdwatchers will lose some old friends. Higher winter temperatures will lead to the loss of cold-adapted species, particularly those that are already on the southern limit of their range. Migrants such as Fieldfares, Redwings, Bramblings and Snow Buntings will be less likely to liven up our dreary winters. Rare but hardy species such as the Ring Ouzel, which loves high mountain summits, are almost certain to disappear within the next few years.

Seabirds, such as Guillemots, Razorbills, Kittiwakes and Puffins are already heading northward to the cooler and more nutrient-rich

In its death throes? The Rathlin Island gull community is struggling to survive.

The Little Egret. A Mediterranean bird that is increasingly at home in Ireland.

waters that are favoured by sand eels, their main food source. The birds that stay will be reduced to living off spiny pipefish, a nutritionally poor food which chokes chicks, stunts growth and can devastate the reproduction of entire colonies, a fate that has already befallen the internationally important seabird community on Rathlin Island.

However, the news is not all bad. Far from it. Twitchers, enthusiasts who would travel the length of the country in search of a glimpse of a rare bird, have already become acquainted with some interesting newcomers. Within the last decade, the Little Egret, a dazzlingly white member of the heron family, common in the Mediterranean and Africa, has become a breeding resident along the south coast of Ireland. Other southern European species, such as the Mediterranean Gull, the Spoonbill and even the Honey Buzzard are also being tempted northward. These handsome interlopers are in the vanguard of what will become an influx of warmth-loving species.[13]

Familiar birds like the blackbird, thrush and sparrow are unlikely to be much affected either way, as they are very widely distributed, successful and adaptable, generalist species.

Insects & bacteria

The most numerous inhabitants of the island of Ireland are of course not humans or cattle, but bacteria and insects. They will be affected in similar ways to our flora, with cold-loving insect species being lost, especially those already at the southern limit of their range. Warmth-loving species will expand their range and abundance. The effects will be mixed. We can expect to see more kinds of butterfly (at present we have just thirty-one). But we will also get more tiny predators and pests. As our freshwaters warm, some temperature-sensitive water bugs will be lost, as will some beetle species. But this must be put in context. It is possible that several dozen kinds of beetle may be lost, out of over 570 rove beetle species known in Ireland.

The effects that rising temperatures will have on organisms that live in the soil, and the alterations that might occur to the soil itself are largely unknown. Common sense, however, suggests that we will have billions of extra insects, as we will have fewer hard frosts to keep their numbers under control. Billions more insects means more food for everything from sparrows to trout and spiders. So the effects of this population explosion are likely to be felt right up the food chain. Nature will of course introduce its own checks and balances, but broadly speaking, the likelihood is that we will have a more numerous and abundant wildlife.

The impact on plant life may be just the opposite. Swarms of extra insects, legions of additional slugs, will eat their way through ever larger numbers of vegetables, cereals and cabbages, pushing our farmers towards the increased use of pesticides. If you think this fanciful, bear in mind the number of entries in the Irish Annals (and other sources) recording insect infestations, including plagues of locusts in 897 and 952, and as late as 1748 in England. Plagues of flying beetles were noted in Connacht in 1688 and 1697, and may yet be recorded again!

Then there is the question of disease. As Ireland warms up, it will become more vulnerable to the arrival of insects and other biological pathogens, which carry diseases. This is of great concern. At present, we have no mosquitoes. But if our average temperatures rise by another few degrees, we can expect to find them living amongst us.

Mosquitoes themselves are bad enough, but with mosquitoes may come malaria or West Nile Virus. If this seems improbable, bear in mind that Canada and the USA have recently seen outbreaks of West Nile Virus which caused hundreds of deaths. More thoroughgoing vaccination programmes, and even sleeping under nets in Cork and

Kerry may also seem fanciful, but in the long term these kinds of minor but telling lifestyle adjustments cannot be ruled out.

Fish (and chips)

Strange things are happening off our coasts. An increasingly wide range of semi-tropical fish and even turtles are being netted in the seas off Ireland.[14] Giant Spider Crabs and Mediterranean Black Sea Urchins have been sighted and caught off the coast of County Antrim.[15] Red Mullet, a Mediterranean species, have become so common off Ireland's coasts that they are being landed commercially for the first time. Red Mullet and chips on a Saturday night? It is no longer the stuff of fantasy.

Even the infamous Puffer Fish, known in Japan as Fugu, is being caught regularly, although we are unlikely to start eating it soon.[16] I ate it once in a specialist restaurant in Tokyo, and it was a hair-raising experience, for the Fugu, a Japanese delicacy, is the second most poisonous vertebrate in the world. Chefs who cook it undergo two years of training. The fish are prepared in a completely separate kitchen area. Their toxic parts are put into a separate bin, which is disposed off as one might dispose of toxic chemicals. The taste is very delicate, but I must admit it took the second flask of saki to give me the courage to order it. Dozens die each year from fugu poisoning, which has no known cure.

Its appearance indicates that the seas around Ireland are becoming more suited to warmth-loving species, and that their balance of species is changing. We will certainly see fewer cod. As our waters warm, the small fish, crustaceans and shrimps that they thrive on (the Atlantic cod is not a fussy eater), will move northwards into cooler waters, and the cod will follow them. This may already have started to happen. It is hard to say. Cod are over-fished, and the dramatic fall-offs in catches may have masked a gradual movement northwards. Either way the prognosis is not good. If fishermen don't get them, climate change will, and in time cod may become an unknown species around our coasts.

The delicately beautiful Arctic Char is also endangered. It can be found in about thirty lakes, ranging from Lough Fad in County Donegal to Lough Inchiquin in County Kerry. Threatened by rising water temperatures and pollution, it has already become extinct in a number of lakes, and is just about hanging on in others. Beloved by

The Atlantic Salmon. Officially considered 'critically endangered' in Ireland. (USFWS)

fishermen, its passing would be much mourned.

Freshwater invaders include the warmth-loving Zebra Mussel, which originated in the Caspian Sea, but has proliferated in Irish lakes. In parts of Lough Erne they cover every rock surface, at a density of tens of thousands per square metre, clogging up water and other pipes, and attaching themselves in vast quantities to boats.[17] The role that climate change has in their explosive emergence in Irish waterways is unclear, but there is no doubt that a warming Ireland offers them an ideal environment in which to proliferate. Our relative isolation from the European mainland has helped to check the flow of newcomers, but in time other species will follow.

A list of Ireland's seven most critically endangered creatures has been compiled. Unlike the endangered plant list, this roster is not made up of niche species. Heading it is the king of fish, the Atlantic Salmon, already near their southern limit and badly hit by over-fishing and the impact of poorly managed salmon farming, which has produced a huge increase in sea lice. Deteriorating early summer and early autumn water quality will hit the salmon during their breeding and hatching seasons.

Also in danger are the Natterjack Toad, the Freshwater Pearl Mussel, the Nore Pearl Mussel, the Twaite Shad, Desmoulin's Whorl Snail, and the Pollan, thought to have been the first fish to colonize freshwater in Ireland at the end of the last Ice Age.[18] Like the salmon, the fish and mussels need healthy freshwater to survive. Their quality of life will be seriously compromised by drier summers and rising water temperatures.

They will also become more vulnerable to pollution given the much lower summer river flows. The Nore and Freshwater Pearl Mussels face an additional danger. Shellfish diseases are likely to become more common because, as water temperatures increase, water quality declines, creating a breeding ground for toxic algae. And the bad news for bathers is that jellyfish are likely to become more prolific as coastal water temperatures rise. They have already made an impression. In 2007 an armada of jellyfish covering an area of ten square miles wiped out Northern Ireland's only salmon farm, killing more than 100,000 fish. Bass, tuna and trigger fish will also find our warmer seas to their liking. We may see more of some of the larger shark species.[19] Great whites have already been spotted off Cornwall. Their arrival here is surely just a matter of time.

9

Human impacts

This chapter looks at how climate change is going to affect our homes, towns and cities, gardens, health, sport and leisure activities, travel and farming.

Our homes, towns and cities

How will climate change affect the way our cities look and function? When the *Sunday Times* asked an artist to produce an impression of post climate change London, he drew Trafalgar Square as a tree-fringed lawn, with wind turbines towering over the famous fountains, and solar panels studding Nelson's Column. White-clad Londoners picnicked on the grass against the background of a windmill filled skyline.[1]

Is this the future? Will the appearance of our townscapes change? Will our houses become festooned with unfamiliar devices? New houses are already fitted with double-glazing, wall and roofspace insulation. But this is just a beginning. In years to come, the search for energy efficiency will do much to shape the character of housing build. Solar panels to heat our water, and photovoltaic panels to convert the energy of daylight into electricity will become *de rigueur* in our houses and apartment blocks. London is already leading the way. It hopes to run 100,000 homes using these devices and wind turbines by 2010.[2]

Communal room and water heating is another – slightly blue sky – option. This is popular in Iceland, where heat from geothermal sources is widely available. In Belfast, Cork and Dublin where densities are high, it may make sense to approach heating communally, rather than have a separate heating system for each dwelling. Inside the home, as electrical devices and gizmos proliferate a parallel effort is underway to limit the amount of electricity used by of some of our more demanding machines. 'White' electrical goods such as washing machines and dishwashers are already rated on

The energy efficient house. (Energy World)

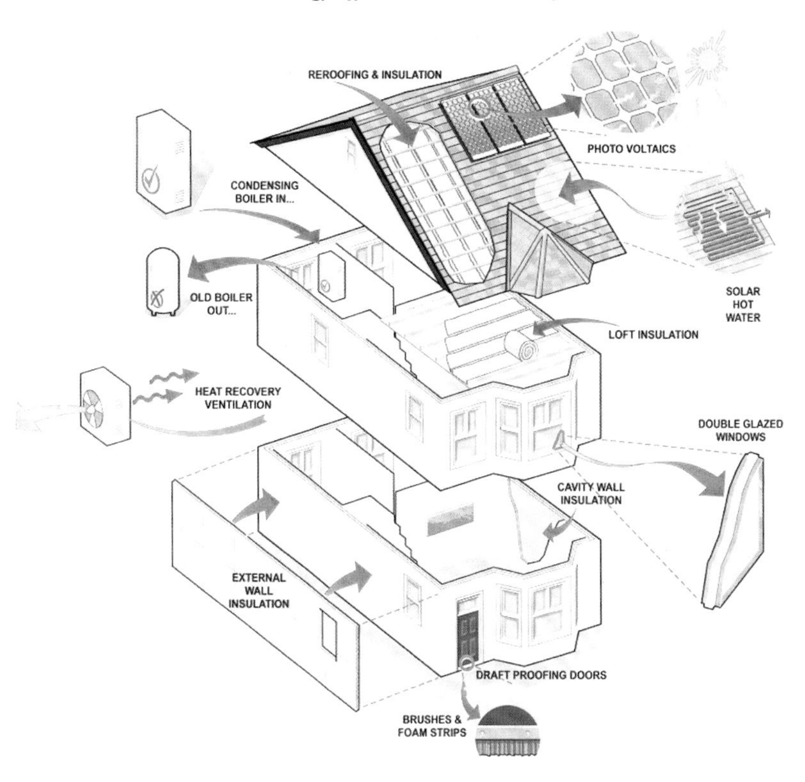

energy efficiency, with a view to making them not just more economical users of energy, but also of water and detergents.

Often the barrier is not technology, but cost. We already know how to build skyscrapers which can glean all the power they need from the sun and the wind. None have been built, however, because they would cost ten times as much as a conventional tower. But the economics are shifting. The Pearl River Tower, currently under construction in Guangzhou, China, will be the world's most sustainable skyscraper to date. It will use 60% less energy than its conventional equivalent, and recoup the additional capital cost in five years. After that it will make a profit.[3]

Here, we are proceeding more modestly. Indeed, we are playing catch-up. During the recent building booms, north and south, large numbers of new dwellings were built to relatively low energy efficiency and conservation standards. It is easy to say now, but this was a missed opportunity. Are our governments doing enough to raise standards? Their record is chequered. In 2007 for example, the Northern Ireland Finance Minister Peter Robinson proposed a rates exemption for everyone purchasing a zero carbon-rated dwelling. In 2008, he rejected amendments to the building regulations which would have required new houses to be fitted with wind turbines, woodchip boilers and solar panels. The south has opted for making the energy efficiency of its housing more transparent. All new houses here are required by law to have an energy rating which will show potential purchasers the 'greenness' of their new home. Standards are rising, but in fits and starts.

Our streetscapes, then, are not about to change beyond all recognition. Things will continue to look much as they do today, with environmental features becoming more integrated within development, as opposed to being add-ons. We will also need less heating. But this saving will be offset by the increased use of air conditioning. Air conditioning is almost unheard of in Ireland, but this will change as the summers get warmer and summer nights become more sultry.

We can expect to see a growth in cafe culture, with an increasing number of restaurants, cafes and bars providing outdoor seating areas, perhaps under canopies, perhaps facilitated by the use of environmentally unfriendly, outdoor burners. The smoking ban has already helped this trend along. Outdoor and patio heaters are

Prost! Warmer temperatures may lead us to embrace a more 'continental' lifestyle. (Belfast City Council)

currently widely tolerated, but the day could come when they are not, especially in a private setting. One environmentally conscious retailer, B&Q, is already committed to ending their sale.[4]

As extreme weather becomes more common, we may imagine storms bringing down increasing numbers of power lines, and our sprawling towns and cities becoming prey to flooding and run off problems of an order that will severely test their antediluvian drainage systems. Torrential downpours delivering two and a half inches (65mm) of rain in one day, which used to occur every fifty years, may occur every ten years. What plans are we making to deal with these and similar contingencies? The August 2008 floods, which turned Belfast's Grosvenor Road underpass into Europe's biggest swimming pool, remind us that we have no reason to be complacent.

Gardens

The drier summers, wetter winters and higher temperatures that lie ahead will combine in fascinating ways in the garden. Lawns will face huge pressures. The longer growing season will demand more mowing and maintenance. Year-round grass and hedge cutting may become the norm. The milder winters will promote weed growth,

Ladas Drive, Belfast, June 2007. A not very happy-looking motorist is rescued. (Press Eye)

which will be hard to deal with for the soil will often be waterlogged and too wet to work. Prolonged dry summers may parch and even kill lawns, as the fescue grasses that are common in our lawns are not well equipped to cope with drought.

Sprinkling or hosing will not always be an option, as water rationing and watering bans are likely in very dry conditions, especially in eastern Ireland. Keeping a water barrel to store run off from the roof of the house will give the gardener a little leverage. Lawns may have to be reseeded with coarser, drought-resistant, Mediterranean-type grasses, such as Bermuda and ryegrass, if they are to cope. But there will be difficulties here too, as these new species will find it difficult to deal with our water-saturated winter soils.

It is not only lawns that will need a fillip. Herbaceous borders will require more intensive care as favourite species like aster, lupin, phlox and delphinium will not adapt well to drier summers. Snowdrops, bluebells, daffodils, and crocuses will not like the additional winter warmth. Tulips, iris and cyclamen will fare better, but will hate the additional wetness. Fuchsia and lavender quite like both and should flourish. Investing in raised beds, and mixing gravel and organic matter with the soil to aerate it and improve drainage,

Sprinkler bans and even water rationing may become a routine part of summer.

will help plants through the waterlogged winter. Cherries and blackberries, which need chilly winters to stimulate flower buds, will find life more difficult. Soft fruits like apricots and nectarines, and even figs and pomegranates, may begin to be grown.[5]

Health

As Ireland is a relatively cold country, a little more heat will in some ways do us all good, at least in the short term, and should help with the incidence of everyday complaints such as colds and flu. Milder winters will cut death rates amongst the elderly. The reduced need for home and office heating will mean less dry skin and fewer respiratory illnesses, such as asthma and bronchitis. Warmer conditions should lead to healthier lifestyles, as we spend more time outdoors and perhaps embrace a more Mediterranean-style diet. More sunshine, more fresh air and more outdoor activity should improve mental well-being and may help to alleviate Seasonally Affective Disorder.[6]

However, it is not all good news. Higher temperatures will lead to more infectious diseases. Increases in various types of fevers can be expected. More respiratory diseases associated with allergic reactions to dust and other allergens could occur as temperature, rainfall and pollution increase, and urban air quality falls. Hay fever sufferers can expect a longer summer allergy season.

Skin cancers will be another, increasing danger. Irish people are amongst the fairest on the planet. Our dominant skin types are unsuited to large amounts of sunshine, and the dangers of increased skin and other forms of cancer are very real. Australian skin cancer rates amongst people of north-western European origin are the highest in the world, four times that of the UK, with a thousand new cases presenting every day, and in excess of 1,400 skin cancer fatalities every year.[7] As my mother, to whom this book is dedicated, and a best friend have both died in the last seven years from secondary cancers derived from skin cancer, I see this as a very disquieting prospect.

Sport and leisure

Climate change will make a difference to our outdoor sports. Team games which are played on grass, such as football, rugby, and gaelic games, will all be affected by the coming changes. In winter, we will see more matches cancelled at short notice, waterlogged grounds, games turned into lotteries by dangerous high winds. In summer we can expect rock hard pitches fast, running games, and nasty accidents. Groundsmen, where there are any, will have their work cut out in trying to ensure that pitches stay green and don't turn into dustbowls during the longer dry spells. Never mind the stress of being a manager or a player. By 2080, if we are still playing on grass, the groundsman's job will be the most demanding in sport.

Some games, however, are just made for the new milieu. Conditions will be perfect for playing and watching cricket. Whether this will be enough to overcome cultural scruples and generate an Aussie or West Indies type cricket mania in Ireland remains to be seen.

The warmer temperatures will mean that outdoor activities such as rambling climbing, sailing and water sports can be carried on for a greater part of the year in relative comfort. Freshwater river fishing, whether coarse or game, will be more of a challenge than ever, thanks to the capricious condition of our rivers, which will be torpid in summer, and in winter all too often in full spate.

Travel

Spiralling oil and diesel prices have already had a huge impact on the way people travel. As petrol prices rose, the most car reliant nation in the world, the USA, recorded a massive drop in car usage, and a proportionate increase in the take-up of public transport. The figures are quite staggering. From January-July 2008, Americans drove forty billion (that's billion!) fewer miles on their highways compared to the same period in 2007. To put this figure into perspective, this is the equivalent of driving from the earth to the sun and back over two hundred times.[8]

High prices have also stimulated the search for alternatives to the conventional petrol engine. The research push is towards low fuel-using vehicles and hybrid, electric and mixed fuel cars. In my own university our gardeners and porters use electric vehicles, adapted from golf carts. 'Smart cars' are also becoming a more familiar sight on the roads. But progress is slow. Public resistance is high. We love our cars, and we are unlikely to give them up until we're offered something better.

Were government to play its hand responsibly, that something better might be public transport. As cities densify and planning increasingly focuses development on train and bus corridors, public transport comes into its own. By encouraging regeneration and getting tough on urban sprawl, government can create the conditions within which bus and train routes become the option of choice, as the success of the Luas light rail system in Dublin has shown. In 2007 it carried twenty-nine million passengers, and is almost working at full capacity. However, even full take up will not allow public transport to break even, and government underwriting of the public transport system will remain essential. In order to maximize the carbon reductions public transport delivers, government should encourage the use of trains and trams, powered by electricity from renewable sources, or from low emission fossil fuels such as gas.

For all the success of trains and busses, we will not be abandoning our cars any time soon. More than half of the south's 1.7 million workers drive to work in their cars, mostly on their own, jamming the roads every rush hour. The number of cars in the north has doubled

Smart car, Belfast.

to over one million over the last twenty years despite only a very small increase in population. And car ownership is increasing. Every year around 100,000 new cars appear on the roads in the south and 50,000 in the north. So if our cars are a given, perhaps we should be a bit more creative about the way we use them, and take a fresh look not just at cars, but at our culture of car use.

There have been some interesting developments in this area. 2008 saw Whizzgo, Ireland's first pay-by-the-hour car club set up in Belfast. For those interested in car pooling, www.carshare.com puts people who wish to car share in contact with one another, be it for regular commutes or occasional long distance travel. Local authorities might also consider promoting bicycle use, as they do in Paris, where, for the payment of a token, bicycles can be taken and returned to bike racks elsewhere in the city. Initiatives to encourage walking, cycling and the use of public transport abound, north and south, but they have not succeeded in reducing greenhouse gas emissions. In 2005-06, transport emissions grew by seven per cent in the Republic, and accounted for thirty-five per cent of energy-related emissions.

Farming in a changed climate

It would be an irony indeed if the human achievement were to be laid waste by – not to put too fine a point on it – a bunch of farting cattle.

It hasn't come to that, of course. It may never come to that. But in emissions terms at least, the humble cow is doing its bit to knock us off our pedestal.

In 2006, agriculture generated 28% of the Republic's greenhouse gas emissions. These took the form of methane expelled by cattle and sheep and nitrous oxide from fertilisers. This figure is down from its 1998 peak of nineteen million tonnes of carbon dioxide equivalents (in 2008 it accounted for an estimated 17.6 million tonnes) thanks to declining numbers of cattle and sheep.[9] This figure is likely to carry on improving as long as stock levels continue to decline.

Animal methane emissions

(kg of methane per animal per annum)

	Wind	Manure
Dairy cattle	100	15.9
Non-dairy cattle	50	6.4
Sheep	8	0
Horses	18	0
Donkeys	10	0
Pigs	1.5	5.4
Poultry	0	0.1

In Australia, which faces the same problem, it was recently suggested that kangaroos, 125 million of them, should replace cattle and sheep, as this would reduce methane emissions by 15% and the resulting meat would be healthier.[10] However, it is hard to see Kangaroo slices pushing rib-eye steaks off Australian dining tables. Or, closer to home, pork chops giving way to ostrichburgers or red deer goujons, even though farming the latter would also be climatically advantageous, because of their lower methane emissions. This rather puts the problem in a nutshell. When climate change clashes with established tastes and preferences, it almost invariably comes out the loser.

Rising temperatures will be both good and bad for farmers. The positives will be fewer frosts, and a longer grass growing season, which will reduce the demand for hay and silage. Grass, though, will be less productive in dryer parts and we might even see some cattle having to be housed in mid summer on occasion. More efficient husbanding of pasture may be able to overcome some of these

Methane makers. Heifers, Portavo.

problems. Certainly farmers will have to rethink issues such as slurry storage, fertilizer inputs, and harvest practices. Higher summer temperatures and dryish grasslands are also likely to hit milk yields, reduce animal fertility, and increase livestock heat stress. New insect borne diseases may become a threat to livestock.[11]

Barley and wheat may do quite nicely in places where summer rainfall reductions are not too marked. Maize, which has become a common enough sight in the Irish rural landscape, will probably fare best of all, and we may see it grown for grain rather than silage. The acreage under biofuel crops like oilseed rape and even elephant grass could also rise.[12] Soya bean may even make an appearance later on in the century, displacing maize.

Around the fringes, we may see novel forms of diversification. Two vineyards were set up in County Cork in 2002.[13] Other vineyards have been set up since, most notably in Wicklow in 2006. Irish vineyards, however, do not have the romantic ambience of their French or German equivalents, as the ripening grapes are usually covered with polythene in order to boost size and flavour.

The dearth of summer rain will mean that water will need to be carefully managed and farmers will have to compete for their share. In 2001, perhaps 15% of Ireland's potato crop was being irrigated, as August rainfall is vital for good yields. This percentage can only increase. And potatoes are not the only crop that will need watered. All our root vegetables, including turnips and carrots, may require irrigation in the medium to long term and, depending upon the availability and cost of water, this could make them unviable.

But the potato may be the biggest casualty. Potato production has dropped by two-thirds over the past twenty-five years, and it will become unviable in many parts by the mid-century, if summer irrigation proves unfeasible. Given the cultural significance of this crop to Ireland, this would indeed be a symbolic event.

Clearly, these developments will shift the whole weight of crop and grain production. Farmers will either adapt or go under, and we will all have to get used to big changes in the appearance of the Irish rural landscape.

10

Government policy and international commitments

This chapter examines the climate change related commitments that have been made by the governments of Northern Ireland (through the UK) and the Republic of Ireland, and there are some significant differences between the two.

These commitments, it can be argued, arise out of top-down and bottom-up pressure. The former comes from international obligations and treaties, which are driven by the ongoing work by a vast array of scientists across the globe. The latter comes from a radical shift in public attitudes towards environmental issues, which has produced a greening of mainstream politics.

Environmental disasters such as the creation of the Antarctic and Arctic ozone holes woke people up to the idea that our actions can have frightening environmental consequences. They also produced flawed, but nonetheless inspiring, templates for international co-operation and gave existing political structures a new sense of purpose. The widening of the ozone holes was halted. People and governments learned that action could be taken on a global scale to successfully resolve environmental issues.

Northern Ireland: could do better

The UK, including Northern Ireland, has obligations under Kyoto to cut emissions of the six main greenhouse gases by 12.5% by 2008-12. In addition, the UK along with other European countries has set itself the goal of reducing carbon dioxide emissions by twenty per cent by 2020, and to achieving a sixty per cent cut by 2050.[1] All these figures are referenced against the 1990 emissions level, the baseline used for all emissions calculations.

In pursuit of these goals, a whole raft of policies and measures have been introduced aimed at reducing emissions across all parts of the UK economy. Fuel charges, grants for low carbon technologies, and grants for conversion to renewable energy sources were introduced. The results have been impressive. The UK as a whole is pretty much on target to achieve its goals, providing us with a model of what strong government intervention can achieve. These commitments were politically bold, for they put social responsibility in front of individual freedom and hit many people in their pockets. But they were based on informed calculation, and the assessment that the greater political pain lay not in acting, but in inaction.

How does Northern Ireland fit into this benign picture? Eighty years ago, it was a considerable industrial force. It had a powerful linen industry, contained the world's biggest shipyard and ropeworks, and had a massive engineering and related infrastructure. It was also a major polluter, in all mediums, as the dank state of the River Lagan testified. But much of this industry has gone, and Northern Ireland now produces a relatively small quantity of greenhouse gases. Most of these come from agriculture, transport and domestic electricity consumption, and as a result, major reductions will not be easily achieved.

Any reductions that can be achieved will help the UK meet its overall targets. According to the latest disaggregated figures, produced in 2003, Northern Ireland only accounted for 3% of the UK greenhouse gas emissions. This figure was 3.5% below the base 1990 level, but it would not have been low enough to meet the Kyoto Protocol target had Northern Ireland been treated separately to the rest of the UK.[2] Northern Ireland accounts for an astonishing 11% of the UK's agricultural emissions, a reflection of the importance of agriculture to the Northern Ireland economy.

In July 2008, the leaders of the G8 group (Japan, Britain, Canada, Germany, France, Italy, Russia and the USA) moved beyond the Kyoto Protocol. They committed themselves to work with almost two hundred of the other countries and territories involved in the UN climate change talks with a view to halving greenhouse gas emissions by 2050. This was an important step forward, because it created a possible framework within which the UN can carry on setting targets once the Kyoto commitments end in 2012, and makes the achievement of a post-Kyoto agreement much more likely.[3]

The reaction of the developing nations to the G8's move was no less interesting. Contrary to what might have been expected, this group, which includes India and China (often portrayed as being lax on climate change), immediately issued a strongly-worded statement criticising the G8's lack of ambition, and advocating a much more aggressive approach towards emission reduction. Another weakness in the G8's position is that it has set no clear path towards target achievement. Intermediate targets will need to be charted out if the 2050 goal is to be reached.[4]

One area in which Northern Ireland lags far behind the rest of the UK and Europe is renewable energy. Northern Ireland has yet to

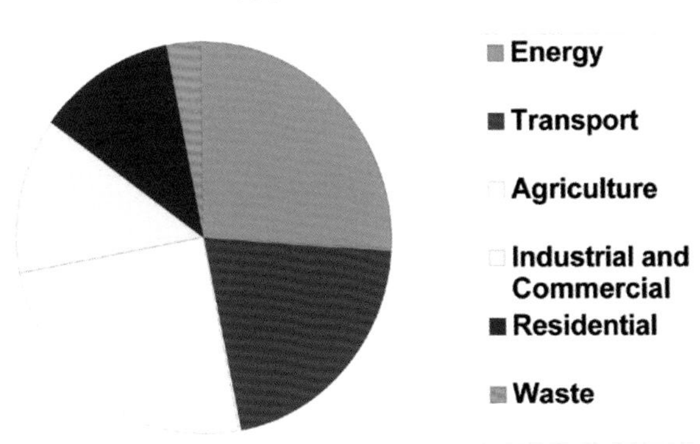

Black north? No longer. Industry and commerce now generate a higher percentage of southern emissions.

invest heavily in the development of renewables such as wind power, and its contribution to the overall UK target is less than *pro rata*. The percentage of electricity which is to come from renewable resources, for example, is set at a lower level here than in England and Wales. In addition, the Northern Ireland target only covers the period to 2012, whereas England and Wales have committed to a target of 25% by 2025.[5] If it is to reach its target Northern Ireland will have to raise the amount of electricity coming from renewables from the current three per cent to 12.5% per cent by 2012.[6] This is hugely ambitious, but necessary if Northern Ireland is to maintain a competitive position relative to the other parts of the UK.

At Kyoto, the European Union committed itself to cutting its emissions by 8% by 2008-2012. It is now anticipated that the fifteen European Union states (including the UK and Ireland) that signed up to this will at least achieve 4% by 2010, and will not fall too far short of the 8% target by 2012. Paradoxically, the current economic slowdown will help them to attain this goal. However, if the new and much tougher target of a 20% reduction by 2020 is to be reached, a much more thoroughgoing set of policies and practices will have to be implemented.[7] As they say, we ain't seen nothing yet.

This will require political commitment. Northern Ireland's pugnacious Environment minister Sammy Wilson is on record as likening climate change to a 'new pseudo-religion', views remarkably similar to those expressed by London mayor Boris Johnston before he read the 2006 Stern Report (Johnston now describes himself as a committed environmentalist).[8] How this will affect Northern Ireland's delivery within the UK target matrix remains to be seen.

The Republic of Ireland: too little too late

The Republic of Ireland is on course to miss its Kyoto targets by a wide margin. Its individual target was generous. It was allowed a 13% *increase* in emissions over and above the 1990 level to allow scope for the development of what was one of the weakest economies in the European Union at that time, and the likelihood of significant growth stimulated by European Union grant money. What was not expected, in any way, shape or form, was that the Republic would grow at a rate of over 6.5% per annum, taking it from being one of the weakest economies in the European Union to one of the strongest.

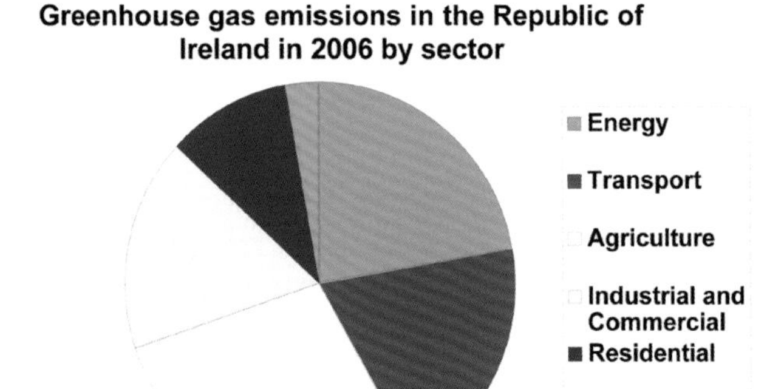

As a result of this phenomenal growth, emission levels have soared. By 2010, Ireland's emissions are likely to be of the order of 25% above the 1990 level, making its 13% target as good as unreachable, even if drastic and politically contentious measures were taken.[9] The global slow down in economic activity brought about by the credit crisis will make the Republic's efforts appear slightly more respectable, but even this will not bring it within striking distance of its 13% goal.[10]

Are the two governments doing a good job?

There is no doubt in my mind that the governments of the Republic of Ireland and Northern Ireland have failed to see the true significance of climate change, and its likely impacts on our environment and way of life. Although progress has been made in curtailing emissions and reducing energy use through recycling, much more needs to be done. Forestry, for example, is an area with immense potential. Trees are ideally suited to the Irish climate and there is a huge demand for wood and wood products across Europe. Because trees draw huge quantities of carbon from the atmosphere, they make a perfect, carbon-friendly product. However, in the Republic, Coillte, the Irish Forestry agency, has virtually stopped new planting and is now reliant on private investment for new plantations.

Both governments could be more pro-active. Only Northern Ireland, through *Preparing for Climate Change in Northern Ireland*,

Forestry: an area in which enhanced grant-aid could do much good. (Belfast City Council)

has systematically assessed the changes that are likely to occur and the possible mitigation strategies required in areas ranging from forestry to health.[11] The sixty-four thousand dollar question now is, does the political will exist to implement it in full?

The Republic has yet to put together a similarly comprehensive strategy to deal with the consequences of climate change. Its National Climate Change Strategy contains broad measures to tackle greenhouse gas emissions but it lacks specifics, which makes it hard to evaluate what has happened and is happening. Vulnerable locations, facilities, infrastructure, buildings, and environments have not been identified. Systematic research into what can best be done to mitigate likely impacts has not yet been undertaken at the scale required to meet the threat of climate change. More funds need to be set aside for this work, and there is no indication that they are likely to be. In short, the north is somewhat better prepared than the south but both are somewhat lacking. Crisis management appears likely to be the order of the day.

Nothing less than a Minster for Climate Change in each part of the island may be needed to co-ordinate our endeavours. Were I asked to write a job description for these ministers, I would suggest that action is needed in four main areas. The first is in the field of engineering and public works, and would involve the creation of river, estuarine, and coastal flood defences.

The second concerns how we manage economic and commercial activity. Here the minister's brief would, in agriculture for example, be to promote farming practices that minimized carbon emissions, and do so without impinging on economic prosperity. This could be achieved by taking greater advantage than we do at present of the growing carbon economy, currently worth around 1.5 billion dollars and rising fast.

The third area would be to do with the way we behave, and would involve creating a culture of carbon neutrality in everyday life – in the kitchen, at the golf club, across the whole gamut of our social and recreational activity.

The fourth area of involvement would be in the less nebulous realms of taxation and public administration.[12] Taxation is perhaps the ultimate carbon killer. But it is a weapon that few governments have dared to use. Bjorn Lomborg, the influential Danish thinker, has suggested that taxing carbon emissions is the surest way to achieve reductions. Two dollars a ton should do it, in his view. Courageously, the Republic's government did move towards introducing a carbon tax, which would have helped the country reach its Kyoto targets. But this was shelved because it would have been politically unpopular, and it may now be introduced in 2010, if at all.

The planning and building control regulations could also be used to positive effect. Some 30% of the Republic's housing stock has been built between 1994-2006. It has been built to what would now be considered relatively poor environmental standards. Sprawling estates have been built on floodplains, but tighter planning policies should cut down the amount of building that takes place in these areas. Northern Ireland's Planning Policy Statement 15 requires flood risk assessments to be undertaken for proposed development within or near to a floodplain.

Flooding could prove a catalyst in more ways than one. The recent floods prompted the Republic's government to announce that it was going to 'bring forward publication' of a climate change adaption strategy.[13] This will make interesting reading, and we must hope that it will do more than gather dust. The extent of the danger can be judged from the fact that Britain considers flooding its biggest

Bringing global warming onto the political agenda. This eye catching image shows a post climate change O'Connell Street.

climate related threat. Flooding in Britain currently costs around £1 billion a year. By 2080 this is expected to rise to a staggering £27 billion.[14]

Ultimately, government's response to climate change is actuarial. It takes the form of an on-going series of cost-benefit analyses. For the plain truth is that, behind the rhetoric of climate change awareness, political decision making is and will continue to be governed by expediency. What will it take for government to make the sea change that science suggests is necessary? Probably a major climatic disaster on this island, involving significant damage and loss of life. Only then will politicians sit up and take notice. Only then will the public be prepared to shoulder the financial burdens involved. Hopefully, by then it won't be too late.

11
What government and business can do

From Bali to Belfast?

At Bali in 2007, the heads of a hundred and fifty global companies nailed their colours to the mast. The evidence for climate change was irrefutable, they declared, and business must adapt accordingly.

Business had not been dragged kicking and screaming to this point. When they looked into the future, the chief executives did not see Armageddon. They saw 'significant business opportunities'. Making the shift to a low carbon economy was the pro-growth strategy. But business could not do this on its own. In the Bali Communique, business leaders called on the world's governments to enter into binding agreements to reduce carbon emissions. This would provide business with the certainty it needed to make serious investments in low carbon technologies. Business was as good as begging government to get its collective act together.

Government's role is crucial. There is much that Ireland's administrations can do, indeed must do if Ireland is to maintain its competitive position in the world. A number of businesses north and south have already recognised this. Responding to changes in public opinion and the regulatory environment, they have redefined where their interests lie, and are in many ways leading the advance towards a more sustainable Ireland.

Water, wind, solar, tidal, biofuels and nuclear power.

One of the key ways in which to build a sustainable future is to promote clean, alternative energy sources. Water power has long been used in Ireland, but always on a fairly small scale. In the south, 5% of electricity comes from hydroelectric power, Ardnacrusha on the Shannon accounting for nearly 40% of this total. By and large, however, our highly developed river valleys are not well suited to damming and hydroelectric power schemes. Our need to guarantee our water supplies will have to become much more acute before the difficulties attaching to proposals for drowning major river valleys will come to be seen as the lesser evil. However, that is not to say

Ireland's first wind farm at Lendrum's Bridge, County Tyrone.
The farm, which was also the first in the UK, opened in 2000.

that small scale schemes should not be considered.

There are some sixty-five wind farms either under construction or in operation in Ireland, north and south, on land and offshore. Not withstanding this, and the huge advances that have been made in the field of turbine generation capacity, the wind remains a largely untapped resource. At present, Eirtricity, a private company which feeds the electricity grids of both territories, operates thirteen wind stations, which generate 280 megawatts of electricity. It has another fifty-six megawatts under construction, including the first phase of the Arklow Bank offshore wind station which, when completed, will be the biggest in the world. Stations capable of generating a whopping 660 megawatts are currently being planned. Ireland's power utility companies, the Electricity Supply Board (ESB) and Northern Ireland Electricity (NIE) are in the van of wind power promotion.

The difficulty the utilities face is not in establishing wind farms, but in integrating them within the existing electricity grids. This is because wind is mostly farmed in the west of the island while the population lives largely in the east. NIE faces this problem in a particularly acute form. Its grid is concentrated in the east of the province, around the main centres of demand and its existing fossil fuel power stations. Its power highways run out in mid-Ulster, after which they become B roads and by-ways, carrying relatively modest amounts of electricity to a mostly rural population. NIE would very much like to harness western wind power, and needs to if it is to meet its renewables target, but it cannot make the connection because the infrastructure isn't there.

Were this Stalinist Russia, making the connection would be a simple matter. The turbines would be built, the pylons would be run up, and that would be that. Renewable energy would flow, unconstrained, into our respective grids. But our system doesn't work that way. It involves complex checks and balances, and although the utilities are chafing to get started (for the large scale integration that is necessary if Northern Ireland and the Republic are to meet their renewable targets will need to be completed within the next ten or fifteen years), local interests and sensitivities will also have to be weighed in the balance. This clash between strategic and local priorities, always a vexed matter, is likely to keep the planners busy for years.

Idyllic looking offshore wind farm.

Seagen's Strangford turbine with Portaferry in the background and (inset) how the turbine works. (Marine Current Turbines)

Power generation, then, is going to become increasingly diverse. As fossil fuel power stations lose their monopoly, and electricity generation becomes more democratic, the utilities will be faced with the challenge of turning our grids from one-way downflows into interactive entities, which will not only allow destinations to receive power, but also allow clean energy from offices, industrial estates and small home turbines to be fed into the grid. This will be a mammoth and very costly operation. In terms of ambition it will not quite match the electrification of Ireland, but it will be a revolution of almost equal magnitude none the less.

Solar power is another potential clean energy source, but its future would appear to lie in the domestic rather than the commercial arena. At present, our electricity grids do not derive any power from solar sources, and the potential for them to do so is low, given the limitations of current technology. Where solar power does come into its own, though, is when solar panels are used to provide effectively free hot water to individual homes and small apartment complexes. For this purpose they are unparalleled, and in the poorest hill villages in Crete, every home has one.

Though we are not as naturally rich in sunshine as Crete, companies such as Genertec Ireland have begun to fit solar panels, which use German technology, to a range of buildings including houses, apartment blocks and factories. These panels are more than capable of heating water, even in Irish weather. But they are far from being a part of every household. We need to make them as common as microwaves and televisions. Government is helping this process along by offering grants to cover part of the cost of purchase and installation.

Surprisingly, given the length of our coastline and the battering we take from the Atlantic, we have scarcely begun to harness wave and tidal energy as sources of power. The difficulties are technical. We have yet to produce machines that are capable of effectively exploiting wave and tidal power, but a number of prototypes are currently being tested in Irish waters, and Northern Ireland has recently become home to the world's first operational tidal power generator, an NIE funded turbine capable of meeting the electricity needs of more than a thousand homes. This experimental device was installed in Strangford Lough in the summer of 2008, and its fortunes are being followed with much interest.[1]

But the kudos does not belong to the private sector alone. Government has also stirred itself. A 1.6 million euro research project, jointly sponsored by the Irish, Northern Irish and Scottish administrations, is currently exploring ways of reducing the regulatory and technological barriers associated with the development of offshore wind, wave and tidal electricity generation installations, with a view to speeding their development, and feeding their energy into national grids.[2] The project will take a couple of years to complete.

Biofuels are a rather more fraught area, and are becoming increasingly so as they are implicated in food price rises and reduced biodiversity. But there may be ways of producing the gains without the pain. One such may be the use of seaweed as a fuel. This is currently being investigated, but it will be some time before its bio-value can be determined. If it proves serviceable, it will add greatly

Seaweed's potential as a biofuel is currently being investigated. (Steven Wooster)

to Ireland's biofuel resources, and sidestep almost all of the difficulties associated with land-based production.[3]

Incineration can also provide additional sustainable energy. It doesn't produce huge amounts and is a bit of a last resort, but it is important in terms of ethos, for it makes clear that we are serious about recycling, the message being 'we waste nothing'. Incinerators currently proposed for Meath and Cork will generate enough electricity for up to twenty thousand houses, and reduce the amount of material going to landfill by 90%, the balance being inert ash. Incineration also 'mops up' items such as furniture and sofas, which are often unrecyclable, and gets some energy back from them rather than dumping them in landfill sites.

Finally, there is the nuclear option. Nuclear power generation produces virtually no greenhouse gases, but this positive is offset by profound environmental concerns – what if there were an accident, and, especially, how do we dispose of the radioactive waste? – concerns which make nuclear power a political non-starter, at least for the present. However, this option has long been favoured by the French, and has recently found an unlikely champion in the form of environmental guru James Lovelock, who has argued powerfully, and without irony, that it is the best way of meeting our energy needs without incinerating the planet.

Integrated resource management – reduce, reuse and recycle
We used to live in a throwaway society but this is gradually changing. In 2007, the Republic recycled a higher percentage of waste electrical equipment than anywhere in Europe. All types of battery can now be recycled, and battery collection points have been installed in shops and elsewhere in response to European Union led programmes on waste control and recycling. Local authorities have also been active in this area. By using 'kerbie' boxes and brown bins, Armagh City and District Council, for example, collects three thousand tonnes of recyclable waste per annum, reducing the amount of material going to landfill by thirty per cent.

However, we still generate huge amounts of waste. Packaging is a case in point, indeed it has become something of a litmus test of our commitment to recycling. Six hundred thousand tonnes of packaging of all descriptions were recycled in the south in 2007, enough to fill the Birds Nest Olympic Stadium. This sounds very creditable, but the fact is that our recycling rates are low, when measured against those of other countries.

Kerbie boxes and guard dog, Belfast.

Gull blizzard. Landfill site on Belfast's north foreshore. (Tom Duddington, Friends of the Earth)

Taiwan recycles a third of all the material it produces, the highest recycling rate in the world. The Netherlands recycles 64% of all domestic waste, and Germany recycles 57%, whereas in 2005 the UK as a whole only recycled 23% of domestic waste. As of 2007, the south had just a 20% domestic recycling rate, but this is rising fast, as is the recycling rate in Northern Ireland. Most of the metal used in Ireland is recycled, and has been for decades. Commercial cardboard packaging is also extensively recycled, glass less so, a shame for it is easily renewed. We are also recycling more wood.

Most of our recycling initiatives are European Union driven. The Repak scheme, for example, would be a good example of this. It is a voluntary initiative between industry and the Department of the Environment aimed at meeting industry's producer obligations under the EU directive on packaging and packaging waste. It achieves this, saving companies money by the by, sums that have increased dramatically as landfill costs have risen.

50% of aluminium is currently recycled, but virtually all of it could be, if we stopped throwing away our fast food containers. (Brazil recycles 87% of its aluminium.) Sadly, although we are recycling more and the range of material that can be recycled is broadening, the increase in our consumption almost offsets this, meaning that the amount of waste that goes to landfill is falling only very slowly. We could do more to push the recycling habit through grants to recycling companies and local authorities, and incentives for the use of recycled materials. So, although we are making progress, it will be a while before we can give ourselves any pats on the back.

We do, however, have one feather in our caps. The south's introduction of a plastic bag tax, an option being considered by many countries including China, was a wonderful innovation. It cut plastic bag use by over 90%, saving at least one billion bags, and the associated bag tax raised over 3.5 million euro for environmental causes. Furthermore, our eyes are no longer confronted by the sight of plastic bag draped trees and hedges every time we travel through the countryside. All in all, this bold scheme has been a resounding success.

Going carbon neutral

Becoming carbon neutral means annulling the carbon we produce by offsetting or 'sequestering' the same amount of carbon. This can be done in a number of ways, including planting trees (which absorb carbon), switching to renewable energy sources, taking steps to reduce carbon emissions, or as an option of last resort, buying carbon credits, a very expensive business indeed.

Talking about going carbon neutral is easy. Doing it is much more difficult, especially for business organisations, most of which will find the objective of significantly reducing their carbon footprint a sufficiently challenging target to be going on with. Getting a grip on direct emissions (the energy used to produce a product or service) is only half the battle. One also has to neutralise indirect emissions, such as the energy consumed by staff as they travel to and from their place of work.

But businesses are making inroads. Musgraves, the food retailer, has installed geothermal heating and solar water heating systems. It also buys its electricity from Eirtricity, thereby reducing its emissions. Glanbia has reduced the carbon footprint of its Botanics shampoo by twenty per cent. Spring Grove, which washes hospital and hotel linen, uses an innovative flash steam capture system to save five hundred tons of carbon annually in its Cork plant.

Remarkably, the first large organisation to become carbon neutral was the Vatican City, which achieved this distinction in 2007 thanks

Greening the workplace. Seppo Leinonen's take on the paperless office.

to the presence of a Vatican owned forest in Hungary, which is used to offset all direct and indirect carbon emissions. However, the Vatican already owned the forest, so it can not exactly rest on its laurels. Samso Island in Denmark, with a population of over 4,000, has also achieved carbon neutrality. It did this by producing wind energy, and introducing an island-wide heating network based on burning biomass fuel.

The GAA is doing its bit. It has pledged to make Croke Park carbon neutral by 2009, which will mean cutting its annual 4,500 tons of direct carbon output. This is being achieved by upgrading the existing electricity, water and waste management systems to the highest possible environmental standards. But it is not just the 'bricks and mortar' that are receiving attention. Supporters around the country are being invited to reduce their personal carbon footprint by pledging to do small things like not overfilling the kettle, which could save fifty kilograms of carbon per person, per year. This initiative is run as an inter-county competition, with the top three counties featuring on the GAA's website.[4]

But this is only one side of the equation. The other involves finding ways of soaking up the carbon that we cannot avoid producing. One of the most efficient and commercially attractive ways to do this is through tree planting. Forestry products are in high demand, and even within the European Union there is a huge shortfall between need and production. As a result, grants are available, north and south, to promote forestry development. The fact that investors in forestry do not pay tax on their returns, adds an additional incentive.

Energy efficiency

The European Union has been the driving force behind all matters relating to energy efficiency in Ireland, and has required both of our governments to promote energy efficiency in every aspect of public, commercial and private life.

Advertising campaigns have advised people about what they can do in the home. The Power of One campaign in the south tried to educate people in using electricity more economically. The most recent television campaign in Northern Ireland featured a dad's attempt to cut the family's electricity bill by fitting insulation and turning off electrical equipment that is left on standby.

It is difficult to quantify the impact of these campaigns, or separate their influence from that of the other influences that bear on peoples' behaviour, not least because the most recent data available on our per capita energy consumption, north (48,648 kilowatts) and south (45,013 kilowatts), comes from 2003. But they may be doing some good. Sustainable Energy Ireland notes that the Republic improved its energy efficiency by just over eight per cent between 1995-2005, which was higher than the European Union average of five and a half per cent. When we look at the figures by sector, we see that transport did poorly with an improvement of under one per cent. Residential users did well, registering an 8.2% increase. Industry did best, recording an improvement of 15%.

Becoming energy efficient is not quite the same thing as embracing environmental responsibility, but industry and commerce are doing this too. Our biggest businesses are positioning themselves environmentally, with a view to obtaining competitive advantage. They compete with one another to demonstrate their environmental credentials. Whether this is all for show, or is the result of a genuine commitment arising from a redefinition of the company's vital interests, or a bit of both, is in many ways irrelevant. The fact is that it is happening, and happening in very creative ways.

Five southern companies – Diageo at their St. James's Gate Brewery in Dublin, Heinz and Xerox in Dundalk, Glanbia in Ballyragget and Abbott Pharmaceutical in Sligo – achieved coveted IS 393 certification in 2008, the first to do so, demonstrating they have achieved best practice in energy management. In the case of Guinness (Diageo), carbon dioxide emission savings of greater than eighty thousand tonnes were achieved between 2004-06 by using a Combined Heat Power plant, partly fired by steam from the brewing process. In 2007, Diageo secured a multi-million euro green energy electricity contract with Bord Gais, meaning that the externally purchased electricity used in its brewing sites across Ireland comes from renewable sources such as wind turbines.

Interesting things are also happening in education. Omagh College of Further and Higher Education uses a biomass boiler to provide all the under floor heating in part of the new campus, dramatically reducing its carbon footprint. The University of Ulster has erected a fifty-five metre high wind turbine at its Coleraine campus. This will reduce campus electricity costs by a quarter and pay for itself in five years.[5] The proposed new library at Queen's University, Belfast, will use the warmth generated in its computer centre to heat the building.[6]

Entering the carbon economy

Businesses north and south are beginning to exploit the opportunities provided by the so-called carbon economy. Northern Ireland's engineering sector is pioneering climate friendly technologies. As well as assembling wind turbines, Harland and Wolff are now developing wave and tidal power generators. Wrightbus, based in Ballymena, is developing fuel efficient hybrid electric busses, fifty of which have just been sold to Nevada for use in Las Vegas.

The Dublin Work Flow Project uses new technology to connect private and work-based computers, giving people the option of working from home as well as in the office. It also provides workers with up-to-the-minute advice on journey times, allowing them to pick the best time to travel to and from work, thereby helping to reduce emissions.

The building group, Carvills, has begun construction of Northern Ireland's first 'eco-village' at Woodbrook, outside Lisburn. A similar village in Cloughjordan, County Tipperary, has just had its infrastructure completed. Residents will now be able to build homes and move in. The village covers an area of sixty-seven acres, forty of which have been set aside for vegetable plots and an orchard.

Valuations of the carbon economy differ, but all forecasts predict strong growth in this sector. The Carbon Trust estimates that the global carbon economy could employ twenty-five million people and be worth five hundred billion dollars by 2050. Even if this estimate were to prove bullish, its value to Ireland will still be measurable in billions of pounds /euro, and tens of thousands of jobs. We must maximise our share of this growing market.

12
What you can do

Getting started

Clearly, there is a pressing need for each of us to act in order to reduce our carbon footprint. Some people say, 'Why bother? Sure it'll make no difference.' I disagree. It makes a difference. The 'why bother' or Chinese power station argument is unacceptably fatalistic. Its logical conclusion is no-one doing anything and nothing getting done – and this is not good enough. Reducing our personal carbon footprint is also the moral thing to do. It is a part of good citizenship. We are passing a heavy burden to our children. We owe it to them to lighten it if we can.

This will mean changing our behaviour. In particular, it will mean changing the way we use energy. If each individual were to use ten per cent less energy, it would have a dramatic impact on our carbon footprint. And the good news is that we don't have to live like monks to do it. Given the profligate way we waste energy, this can be achieved without greatly affecting the quality of our lives.

According to the European Union, households generate 16% of Ireland's greenhouse gas emissions, or eleven tons of mostly carbon dioxide per person, per year. Of this, 61% comes from general electricity usage and heating in the home, and 21% from transport, both of which are heavily reliant on fossil fuels, in particular oil.[1] And given that the 150 or even 200 dollar barrel of oil is no longer the stuff of fiction, it makes sound financial sense for us to cut our usage too.

Much of what follows will be mind-numbingly familiar. But no self-respecting work on climate change can be considered complete without it. For it is not enough just to inform ourselves about global warming, we also need to act. If talk like that puts you on your guard, and you are already cringing in anticipation of a finger-wagging sermon, don't worry, there isn't one coming. Just some (hopefully) helpful advice.

Energy use in the home

Great savings can be made in the home. If your house is young, and fitted with proper insulation and double glazing, then you already have a head start. Older houses can be more difficult to work with, but they can also achieve larger gains. The savings, set out in italics below, are calculated in terms of 'kilograms of carbon dioxide per household per year',[2] but don't forget about the cash savings, for they are just as attractive.[3]

Heating

Lagging the immersion tank – *saves up to 330kg*
Insulating the loft – *saves 350kg*
Adding cavity wall insulation – *saves 1,000 kg*
Double glazing – *saves 920kg*
Photovoltaic panels – *save over 1,000 kg*
Solar hot water panels – *save 460kg*
Draft proofing and closing windows and doors – *saves up to 350kg*
Installing individual room thermostats – *saves 600kg*
Fitting a modern boiler – *saves 900kg*
Replacing electrical heating with a woodchip burner – *saves 5,000kg*

If you are very ambitious you might even investigate the pros and cons of installing a small wind turbine. This would not only reduce the household's carbon footprint, where the electricity can be fed into the grid, it will save or even make the owner money. Using heating and electrical items efficiently can also noticeably reduce energy use.

Insulated hot tank, with enough leakage to keep the airing cupboard warm.

For example, a one degree centigrade reduction in temperature can save up to 300kg of carbon dioxide per household per year, and cut 5-10% from the energy bill. Switching off electrical equipment like computers and televisions when no-one is using them can make a difference. Using an outdoor clothes line, as opposed to a tumble drier, will also save pennies and emissions.

Machines & appliances
Boiling only the water you need when using the kettle – *saves 50kg*
Using energy efficient light bulbs – *saves 30kg*
Not leaving electrical equipment on standby – *saves 190kg*
Switching off unnecessary lights – *saves up to 145kg*
Running the dishwasher and washing machine only when full – *saves 20kg*
Running tumble driers only when full – *saves 78kg*
Buying the highest energy rated electrical appliances – *saves 27kg*

Leaving your mobile phone charger plugged in and switched on when there is no phone charging uses three-quarters of the electricity you use while charging.[4] The fridge and freezer work more efficiently when full. Filling the empty spaces with newspaper sounds a little crazy, but it will help them to work better. Siting the fridge and freezer away from hot machines like the cooker or the boiler will also help.[5]

Phone chargers use electricity, even when they're not charging.

Covering pots when cooking saves energy, as does keeping rooms warm by closing doors. It's all about not being wasteful.

Our water use is is also worth looking at. Most of the 'fresh' water that comes into the home has passed through a water treatment plant. This means that every drop of water we use has an associated energy cost and hence a carbon footprint, as does the waste water leaving the home, so the less water we use, the less energy we consume. Having a shower instead of a bath is a simple way of reducing water and energy use, without comromising on hygeine. The installation of a two-flush toilet, common on the continent, will also help. Leaving taps running can also use up a surprising amount of water and hence energy.

Then there's recycling. Its benefits are well known. All glass, paper and cardboard can be recycled, as can most metal and plastics at highly efficient energy levels. Beer and soft drinks cans are the perfect example, for it takes ten times as much energy to make an aluminium can as it does to recycle one.[6] Where possible, electrical goods including mobile phones, which tend to be replaced on a fashion basis, should be recycled, and where possible damaged items should be repaired instead of replaced. Again, it's all about common sense and keeping consumerist over-indulgence at bay.

The food shop

The way we buy food is also worth looking at. Using one of the handy, new re-usable shopping bags, buying items with low transport miles, getting locally grown fruit and vegetables when they are in season, or even joining an organic vegetable and fruit box scheme – if you can find one – are all positive steps. Buying goods that aren't swathed in packaging, buying refills, and purchasing goods which are made from recycled paper, like toilet paper, are also helpful. Minimalism is the key. And this needn't make shopping boring. There is a lot of satisfaction in getting what you want, in the form that you want it. And by doing so, you may even encourage producers out of their old, environmentally careless ways.

In the garden

If you have a garden, why not set a small part aside for growing your own fruit or vegetables? And if the experiment works, try bringing a little more ground into cultivation. (Think of the mowing you'll save!) Maybe your upper limit is a window box. These are perfect for a selection of herbs. This not only provides fresh and healthy food but cuts down on air miles and outgoings, very welcome given the rising price of food. And of course nothing tastes better than your own freshly picked fruit or veg.

Have you the room to plant a tree? Planting a tree can take sizeable amounts of carbon out of the atmosphere, as well as soaking up pollutants such as PM 10s, which are produced by vehicle exhausts and are also implicated in asthma and other respiratory conditions.

Something to aspire to. Carrots, leeks, parsley, cabbage, celery and onions flourish in this vegetable garden. (Steven Wooster)

The amount that each tree 'eats' will depend on the species, but by way of a rough measure, every five trees planted will consume around a ton of carbon. Or you could go one better and plant a fruit tree. You'll not only beautify your garden, you'll have your own food, save on bills, and reduce the food miles associated with transporting fruit long distances.

If you can also compost your food and organic waste, so much the better. It will fertilise the garden, doing wonders for the lettuce and carrots while at the same time reducing the amount of waste that goes to landfill. If you really get into it, you can also use a wormery to recycle your food waste into compost for the garden.

In the workplace

In the workplace, as in the home, there is a lot of potential for energy saving. These savings can come from things like making buildings more energy efficient, through to the use of recycled paper in the photocopier, photocopying on both sides of the page, recycling toner cartridges, reusing paper that has only been printed on one side for rough work, and using ceramic cups and plates in the canteen (instead of disposable ones). Even small changes in large work environments can produce useful savings. Having rooms and corridors on dimmer switches and only heating rooms that are in use can save appreciable amounts of energy.

Savings can also come from questioning established practices and routines. Can we share our journeys to work? Do we always need to work in the office? Working from home cuts out transport energy use, and leads us to more carefully consider the way we use our offices and buildings. Likewise using teleconferencing for meetings, as opposed to sending people hither and thither. Improvements in IT should make this an increasingly viable option.

Transport and travel

We will not easily be weaned off our cars. They are too comfortable. Too convenient. That's accepted. But do we over-use them? Could we sometimes walk to the local shop or takeaway, instead of driving? Could we get the bus to work, and save the car for the trips to B&Q or Tesco? The carbon savings involved in making these lifestyle changes have been reliably quantified in terms of savings of carbon dioxide per household per year over average commuting distances, and they make interesting reading:

Walking to work – *saves up to 200kg*
Giving someone a lift – *saves 15kg*
Cycling to work – *saves up to 350kg*
Taking a bus instead of driving to work – *saves 575kg*
Taking a train instead of driving to work – *saves 720kg*
Car pooling – *saves 700kg*
Cutting out flights and holidaying closer to home – *can save 1,000kg*

Having said this, I should confess that I can lecture no-one on this subject. I live six miles from my work and the bus service is poor, so I almost always get there by car. I compensate by trying to cut down on trips, and by going to several places – maybe work, the supermarket and the library – every time I take out the car. Not using a wasteful or an almost empty car is another sensible step, as is using a car with a small engine. Running a suburban 4x4, on the other hand, should be made a criminal offence, or the next thing to it.

Using a hybrid car, or a car that partly runs on biofuel would also help enormously. Saab, for example, have just launched a range of cars which can use high levels of biofuels, potentially reducing their emissions by up to 80%. However, biofuels are not just for the well-to-do, they are for everyone. Most modern diesel cars can use up to

How sensibly do we use our cars?

a 5% biofuel mix without alterations. In Ireland, biodiesel is made from waste vegetable oil, which would otherwise end up in landfills, and its use can reduce emissions substantially.

There are many things that we can do to reduce car fuel consumption. One of the most direct ways to do this is by buying a fuel-efficient car. Cars with an A or B energy rating pay less road tax and can save a thousand kilograms of CO_2 per year, making them cost efficient as well as climate change friendly.[7]

Correct tyre pressures, low viscosity motor oil, removing the roof rack when it is not required, and following recommended cruising speeds all help the car to run more efficiently. Maintaining a steady speed increases fuel efficiency. Washing your car by hand, or getting the kids to do it instead of using a car wash also saves energy.[8] Many of the things that we can do are not that difficult. And although they may sound new fangled and modern, they aren't really. Carbon-awareness has breathed new life into traditional values and virtues such as economy, responsibility and thrift. And that is maybe no bad thing.

13

What will happen if we don't act now

This chapter looks at what could happen if we fail to control our carbon emissions. Although the following scenarios may seem implausible and even alarmist, all are based on relatively uncontroversial projections of present trends. What is perhaps most frightening about them is their telescoping time frame. Many eventualities which were thought to be thousands of years distant in the Third IPCC Assessment of 2001[1] have been brought forward, in some cases to just hundreds of years off, in the Fourth IPCC Assessment of 2007.[2]

This has come as quite a shock. Scenarios which one could complacently ignore have suddenly acquired a horrible clarity and presence. Six additional years of research, and more importantly, the accelerations in the scale and rate of change that have occurred between the two reports, have brought everything closer.[3] But not close. The four scenarios outlined below are possible eventualities only. All draw on the material presented in earlier chapters. Though interlinked, they are presented singly, as it is easier to understand their impact if they are considered in this way.

Runaway global warming

At present, the best guess prediction is a rise of around 3ºC in global temperature by 2100, well above the critical 2ºC threshold. If we fail to act, the upper limit prediction of around 4ºC will be reached or exceeded.[4] Warming on this scale over such a short time frame would be catastrophic for the earth and its inhabitants. It would have a dramatic impact on the planet's landscape and vegetation, and in turn, on agriculture and our ability to feed and support ourselves. We would see increased famine and starvation, death due to drought, war over food and water resources, and migration on an unprecedented scale, leading to a huge and a probably unmanageable global refugee crisis.

Mortality would significantly increase. Hundreds of millions of people would die, leading to a sharp reduction in the global population. The comfortable lifestyle that the developed world enjoys would be relentlessly degraded, as healthcare systems became unsupportable, global trade crumbled, more and more states became unviable and international stock markets collapsed. There would be mass extinctions of plant and animal species.

Ireland's people would see horrors beyond their boundaries. At home, they would be faced with a drying landscape, which would necessitate radical changes in farming. Many native plant species would disappear. So would some of our bird species. Our boglands would dry out. There would be huge pressure on water resources, and a lengthy roster of water use regulations to be observed. Economic migrants from Mediterranean Europe and perhaps beyond would seek admission.

Is this projection perhaps too gloomy? Will we not, in time, find a way of containing carbon levels? Many potential correctives have been mooted, most of them based on intensifying natural modes of carbon capture. No less a figure than James Lovelock, the creator of the Gaia hypothesis, has suggested the mass placing of very long

Algae growth can be artificially boosted by the presence of iron filings.

(two hundred metre) tubes in the world's oceans. The idea is that these would draw cold water to the surface, stimulating the growth of algae, which absorb carbon. The idea has been taken up by an American company, which is currently trialling them. Another American company, Planktos, is attempting to capture carbon using huge quantities of iron filings, which it dumped into a carefully chosen corner of the Pacific to produce a boom in the growth of algae.[5]

Others have suggested taking CO_2 out of the atmosphere using giant cooling towers containing sodium hydroxide; or through the use of large, chemically treated 'air scrubbers', which are shaped like giant fly swats. The idea is that both would neutralise carbon, which would then be stored underground in old mineshafts.[6] A small, pilot storage scheme is currently underway in Norway, and a study into the feasibility of this kind of carbon sequestration is about to take place in the Irish Sea.

This is all well and good, but it is not a solution. Lovelock's tubes may work stunningly well, but even if they did and proved durable, well over a hundred million of them would be needed to begin to capture meaningful quantities of carbon. Some scientists are unhappy with this whole approach. They argue that the search for so called 'silver bullets' concentrates the debate at the wrong end of the spectrum, and is essentially a form of escapism which lets the known offenders off the hook. But that is another argument entirely. The brutal fact is that, at this point in time, there is no credible technological answer to the problem of excess CO_2. At the moment we are producing toys. Creating processes that are equal to the colossal and global nature of the problem is another matter entirely.

Runaway global sea-level rise

The second scenario, which is in effect part of the first, is that runaway temperatures will produce a rise in global sea-levels on a scale unknown since the end of the last ice age. As we know, the current best estimate is that we shall experience a 44cm sea-level rise by 2100. This assumes no significant ice melt. However, the polar regions are warming at twice the rate of the rest of the planet, and as we have seen in chapter five, there is increasing evidence that ice melt is becoming a factor to be reckoned with.

Polar sea ice, for example, is melting so rapidly that the Arctic seas may become ice-free in summer in under ten years, a development that, a few years ago, was thought unlikely to occur until 2080.[7] The

Ice is melting at a faster rate than anticipated. (Martha Holmes)

The archipelago formerly known as Ireland. This is what would remain of Ireland if the polar ice caps melted. (Chaosheng Zhang)

1,800,000 square kilometre (700,000 square mile) Greenland Ice Sheet is melting at the rate of 257 cubic kilometres (62 cubic miles) per annum.[8] By 2080, the annual melt rate is expected to rise to 465 cubic kilometres (111 cubic miles). If all the ice on earth were to melt, global sea-levels would rise by some eighty to eighty-five metres. This is extremely unlikely, even on a timescale of hundreds of years. However, were we to see a five metre increase, as a result of runaway temperature rise and a correlative ice melt, the consequence would be global coastal flooding and coastline retreat, and the loss of enormous tracts of land in countries such as the Netherlands and Belgium, unless their sea defences were substantially strengthened, a measure that could prove so costly that these nations would have no choice but to abandon large rafts of land.[9]

The world's great coastal cities would be flooded.[10] River deltas would become vast estuaries, with the loss of the numerous cities that occur on their banks, such as New Orleans, Shanghai and many others. There would be increased vulnerability to storm surges associated with major storms and hurricanes. Many of the world's most fertile river valleys, and such cradles of civilisation as the valley of the Nile and the lower parts of the valleys of the Tigress and Euphrates would be flooded; all of which would put additional pressure on food production, and compound the problems outlined above.

In an Irish context, even a five metre rise in sea-level would effectively flood the centres of most of the island's major cities and towns, including Dublin, Belfast and Cork, cities which already have significant flood concerns. There would be wholesale loss of land along the east coast of Ireland in particular. All of our beaches would disappear, and low-lying locations along the entire coast of Ireland would be inundated. Inland, rising water tables would create new lakes and ponds, and water-logging would impede soil fertility. The island would lose some of its best agricultural land, compromising what ability it has to meet its own food needs.

Switching off the Gulf Stream

The Gulf Stream currents have weakened considerably over the last fifty years. Most computer models show them weakening further by 2100, but remaining active.[11] The partial melting of the Greenland ice sheet could, however, shut them down by releasing huge volumes of fresh water into the North Atlantic, which would 'switch off' the thermohaline circulation, stopping the Gulf Stream in its tracks. This would plunge Europe into coldness.

The most disconcerting aspect of this scenario is that it has happened before, most recently during a period called the Younger Dryas, some 12,800-11,500 years ago. This led to a dramatic collapse in European temperatures, with icebergs ranging as far south as the coast of Portugal. Temperatures fell by 5-10°C in just a couple of decades – the merest blink of an eye in climatic time – after which cold set in for the best part of seven hundred years.[12]

Were the Gulf Stream to stop flowing, conditions in Ireland would be comparable to those found in parts of Northern Canada today. Ice, frost, and sub-zero temperatures would be the norm in the cold half of the year. Farming, to pick just one aspect of life, would become immeasurably more difficult. There would be dramatic declines in fertility and yields. Grass and other crops would have a very short growing season. Farmers would have to move from planting wheat towards cultivating sturdier plants such as rye and oats, the kinds of cereals that are grown in Scandinavia today. Huge amounts of additional energy would be required to grow what we need. The prices of foodstuffs produced in this way may be imagined.

This might not be the end of the sequence. It has been credibly suggested that the shutting down of the world's thermohaline currents could trigger a new ice age.[13]

Opening the methane vaults

Peat bogs are not, of course, unique to Ireland. They are widely distributed across the northern regions of the world, where they cover an area in excess of one and a half million square miles, or roughly forty-five times the size of Ireland. (There is a bog in Siberia which is bigger than Ireland.) As bogs are made up of semi-decayed plants, they store huge amounts of methane, an extremely potent greenhouse gas.

Because most of these methane vaults are sealed by permafrost, they make no contribution to the build up of atmospheric greenhouse gasses. However, recent temperature rises and rainfall changes have begun to thin their permafrost caps. This is a serious matter, for if these gasses were to be freed they would push temperatures well beyond predicted levels, bringing about potentially catastrophic global warming.

Bogs are not the planet's only reservoirs of methane. Large quantities are also stored as nodular gas hydrates in the sediments that make up the ocean floors. These hydrates have been identified as a possible source of global warming. Were they to again become gaseous, as a result of rising ocean temperatures, the effect would be to raise atmospheric methane levels, further intensifying global warming. The potential impacts have not yet been quantified, as our understanding of the problem is in its infancy. Suffice to say that it is a cause of concern.[14]

Conclusion

Our course of action is clear. We must seek to control greenhouse gas emissions in order to slow the pace of climate change. We must act in every sphere of life, from the individual to the collective, from the local to the global. Time is short – the next twenty years will be critical, and we may not even have that long.

If we fail, the planet, though tested, will survive. But whether we will play any significant part in its continuing story remains to be seen. It is a mistake to believe that the world is ours. We are not the earth's owners, we are its guests. And like guests who have tried the patience of our host, we may be about to be shown the door. At the very least, human society may be about to be plunged into a period of crisis, a Dark Age that will make the last look tame indeed.

We have only this place. It is time we looked after it.

Notes

Chapter 1

1 Ahlstrom D. 'Why Climate Change Denial Persists', *Irish Times,* 28.8.2008

2 Lomborg B. (2007) *Cool It: The Sceptical Environmentalist's Guide to Global Warming*, London, Marshall Cavendish.

3 McCarthy M. and Frawley M. 'Ireland Second-Safest from the Effects of Climate Change', *Sunday Tribune*, 6.7.2008.

4 Within the last decade, four islands in the Sunderbands between the Ganges and Brahmaputra, have been permanently flooded, making 6,000 families homeless. A number of coral atolls including Tebua, Tarawa, and Abanue in Kiribati, some of the Marshall Islands in the Pacific Ocean, and of course the Maldives in the Indian Ocean, have gone under, and some of the low-lying islands of Vanuatu have had their populations evacuated because they will soon be permanently flooded. Many of the remaining islands in the groups mentioned above, along with other island groups, have experienced severe erosion and flooding as a result of sea-level rise and will soon join the growing list of casualties.

Chapter 2

1 Dixon et al. 2001.

2 Redfern 2000.

3 Robock 2000.

4 Harrington 1992.

5 Carr 2003.

6 Galvin (forthcoming).

7 Danny Harvey 2000.

Chapter 3

1 Britton 1937, p.9, 16.

2 Widell 2007, p.54.

3 Baillie 1995.

4 Britton 1937.
5 Baillie 1995.
6 Britton 1937, p.49, 58, 73.
7 Dawson et al. 2007 Dawson and his colleagues found high concentrations of salt in the ice core deep within Greenland, far from shore. The only possible source of these could have been the North Atlantic.
8 Ross 1992.
9 Smith 1774.
10 Tuckey 1837. The record doesn't clarify which.
11 Dickson 1997, p.12.
12 Ibid. p.51.
13 Carr 1993.

Chapter 4
1 McGuire B. (2002) *A Guide to the End of the World*, Oxford University Press.
2 Manley G. (1974) 'Central England Temperatures: monthly means 1659 to 1973', *Quarterly Journal of the Royal Meteorological Society*, 100, 389-405.
Parker D.E., Legg T.P. and C.K. Folland C.K. (1992) 'A new daily Central England Temperature series, 1772-1991', *International Journal of Climatology*, 12, 317-342.
3 Intergovernmental Panel on Climate Change (IPCC) (2007) *4th Assessment Report;* Working Group 1 Report 'The Physical Science Basis', chapters 3 and 10.
4 Ibid.
5 Ibid.
6 Ibid.
7 Ibid.
8 When the Irish and global ten year running means are compared, Ireland's temperature is generally higher. Since 1991, Ireland has been warming twice as fast as the global anomaly.
9 Met Eireann (2007) 'Warmest year on record in places: sunnier than normal everywhere and dry in south', *Monthly Weather Bulletin,* no. 260, 16.
10 The Community Climate Change Consortium for Southern Ireland Project's objective is to conduct climate change research, develop regional climate modelling capacity, and provide climate model output to Irish scientists. It focuses on generating projections of Ireland's climate. Met Eireann is one of its major contributors.
11 Fealy R. and Sweeney J. (2008) 'Statistical downscaling of temperature, radiation and potential evapotranspiration to produce multiple GCM

ensemble mean for a selection of sites in Ireland', *Irish Geography*, 41 (1) 1-27.
12 Ibid.
13 Kosatsky T. (2005) 'The 2003 European Heatwave', *EuroSurveillance*, 10 (4) ii-552. A temperature of 100.2˚F was recorded at Heathrow, 100.6˚F was recorded in Gravesend, Kent.

Chapter 5

1 Holgate S.J. (2007) 'On the decadal rates of sea-level change during the twentieth century', *Geophysical Research Letters*, 34, L01602, doi:10.1029/2006GL028492.
2 Molnia B.F. (2006) 'Alaskan Landscape Evolution and Glacier Change in Response to a Changing Climate', *Weather*, 61 (3), 84-88.
3 Abdalati W. (2006) 'Recent Changes in High-Latitude Glaciers, Ice Caps and Ice Sheets', *Weather*, Vol. 61 (4), 95-101.
4 Intergovernmental Panel on Climate Change (IPCC) (2007) *4th Assessment Report;* Working Group 1 Report 'The Physical Science Basis', chapters 5 and 10.
5 McGuire B. (2002) *A Guide to the End of the World*, Oxford University Press.
6 *ID21 Insights*, issue 71, January 2008, editorial.
7 Intergovernmental Panel on Climate Change (IPCC) (2007) *4th Assessment Report;* Working Group 1 Report 'The Physical Science Basis', chapters 5 and 10.
8 Measures of 'relative sea-level' change are obtained by referencing sea-level against a fixed, land-based Ordnance Survey level.
9 Carter R.W.G. (1990) 'The Impact on Ireland of Changes in Mean Sea-Level', *Programme of Expert Studies on Climatic Change*, No. 2, Dublin, Department of the Environment, Government of Ireland.
10 Woodworth P.L., Tsimplis M.N., Flather R.A. and Shennan I. (1999) 'A Review of the Trends Observed in British Isles Mean Sea Level Data Measured by Tide Gauges', *International Journal of Geophysics* 136, 651-670. Belfast data collection period 1918-63, margin of error plus/minus 0.34mm per year.
11 Ibid. Malin data collection period 1959-94, margin of error plus/minus 0.68mm per year.
12 Ibid. Dublin data collection period 1938-96, margin of error plus/minus 0.30mm per year. These trends have recently been confirmed by Orford et al., 2006 *Phil Trans Roy Soc*, A. 364, 857-866.
13 There is no published data for any more southerly tide gauge.
14 Sweeney K., Fealy R., McElwain L., Siggins L. and Sweeney J. (2008)

Changing Shades of Green: The Environmental and Cultural Impacts of Climate Change in Ireland, Irish American Climate Project, Berkeley, USA.
15 Carter R.W.G. (1990) 'The Impact on Ireland of Changes in Mean Sea-Level', *Programme of Expert Studies on Climatic Change*, No. 2, Dublin, Department of the Environment, Government of Ireland;
Boelens R.G.V., Maloney D.M., Parsons A.P. and Walsh A.R. (1999) *Ireland's Marine and Coastal Areas and Adjacent Seas: An Environmental Assessment 1999*, Dublin, Marine Institute.
16 Andrew Cooper, Director, Coastal Research Group, University of Ulster, speaking in 2002.

Chapter 6

1 Intergovernmental Panel on Climate Change (IPCC) (2007) *4th Assessment Report;* Working Group 1 Report 'The Physical Science Basis', chapters 3 and 10.
2 McElwain L. and Sweeney J. (2007) *Key Meteorological Indicators of Climate Change in Ireland*, Wexford, Environmental Protection Agency, Ireland.
3 Intergovernmental Panel on Climate Change (IPCC) (2007) *4th Assessment Report;* Working Group 1 Report 'The Physical Science Basis', chapters 3 and 10.
4 McKenna's *English-Irish Dictionary* (1935). With thanks to Aodan Mac Poilin of the Ultach Trust, Belfast.
5 McElwain L. and Sweeney J. (2007) *Key Meteorological Indicators of Climate Change in Ireland*, Wexford, Environmental Protection Agency, Ireland.
6 Sweeney K., Fealy R., McElwain L., Siggins L. and Sweeney J. (2008) *Changing Shades of Green: The Environmental and Cultural Impacts of Climate Change in Ireland*, Irish American Climate Project, Berkeley, USA.
7 Ibid.
8 Ibid.
9 Ibid.
10 Wang S., McGrath R., Semmler T., Sweeney C. and Nolan P. (2006) 'The Impact of the Climate Change on Discharge of Suir River Catchment (Ireland) Under Different Climate Scenarios', *Natural Hazards and Earth System Sciences,* 6, 387-395.
11 Northern Ireland Water 2007/2008 Summary Annual Report, www.niwater.com.
12 In 2007, Dublin embarked on a major water pipe replacement programme. It is hoped that this will cut leakage substantially, but it will by no means eliminate all losses.

Chapter 7

1 Intergovernmental Panel on Climate Change (IPCC) (2007) *4th Assessment Report,* available online at www.ipcc.ch. Look also at the Working Group 1 Report 'The Physical Science Basis', especially chapters 3 and 10.
2 Alexandersson H., Tuomenvirta H., Schmith T. and Iden K. (2000) 'Trends of Storms in NW Europe Derived from an Updated Pressure Data Set', *Climate Research* 14 (1) 71–73.
3 Quinn W.H. and Neal V.T. (1995) 'The Historical Record of El Niño', in Bradley and Jones (eds) *Climate Since AD 1500*, Routledge, 623-648.
4 Intergovernmental Panel on Climate Change (IPCC) (2007) *4th Assessment Report;* Working Group 1 Report 'The Physical Science Basis', chapters 3 and 10.
5 The Armagh weather diary covers the period 1796-2002, Edinburgh covers 1770-1988, Valentia covers 1900-2006.
6 Hickey K.R. (2003) 'The Storminess Record from Armagh Observatory 1796-1999', *Weather*, 58 (1) 28-35.
7 Hickey K.R. (in press 2009) 'The hourly gale record from Valentia Observatory, SW Ireland 1900-2006', *Journal of Climatic Change.*
8 Dawson A.G., Hickey K.R., McKenna J. and Foster I.D.L. (1997) 'A 200-Year Record of Gale Frequency, Edinburgh, Scotland: Possible Links with High-Magnitude Volcanic Eruptions', *The Holocene*, 7 (3) 337-342.
9 Ibid.
10 McGrath, R., Nishimura, E., Nolan, P., Semmler, T., Sweeney, C. and Wang, S. (2005) *Climate Change: Regional Climate Model Predictions for Ireland*, Environmental Protection Agency, ERTDI Report Series 36.
11 Intergovernmental Panel on Climate Change (IPCC) (2007) *4th Assessment Report;* Working Group 1 Report 'The Physical Science Basis', chapters 3 and 10; Dawson A.G., Hickey K.R., Dawson S., Elliott L., Foster I.D.L., Wadhams P., Holt T. Jonsdottir J., Wilkinson J., Smith D.E. and Davis N. (2002) 'Complex North Atlantic Oscillation (NAO) index signal of historic North Atlantic storm track changes', *The Holocene*, 12 (3) 363-369.

Chapter 8

1 Norton B.G. and Ulanowicz R.E. (1992) 'Scale and Biodiversity Policy: a Hierachical Approach', *Ambio* 21 (3) 244–249.
2 Bullock C., Kretch C. and Candon E. (2008) *The Economic and Social Aspects of Biodiversity: Benefits and Costs of Biodiversity in Ireland*, Department of the Environment Heritage and Local Government, Dublin, The Stationery Office, especially Ch.10 Biodiversity and Climate Change.
3 Berry P.M., Dawson T.P., Harrison P.A. and Pearson R.G. (2002)

'Modelling Potential Impacts of Climate Change on the Bioclimatic Envelope of Species in Britain and Ireland', *Global ecology and biogeography* 11, 453-462.
4 O'Keeffe C., Lynn D., Weir G., Fernandez Valverde F. and Roller J. (2008) *The Status of EU Protected Habitats and Species in Ireland*, National Parks and Wildlife Service, Department of the Environment, Heritage and Local Government, Government of Ireland.
5 Sweeney J., Donnelly A., McElwain L. and Jones M. (2002) *Climate Change Indicators for Ireland*, Wexford, Environmental Protection Agency
6 Wyse Jackson P. S. (2007) 'The Potential Impact of Climate Change on Native Plant Diversity in Ireland', *Botanical Gardens Journal* 4 (2) 26-29.
7 Ibid. These figures are based on a mid-summer temperature rise of 2.5 degrees centigrade, currently anticipated in c.2080 (see chapter 4, Table 4.1)
8 Under the terms of the 1985 Wildlife Order in Northern Ireland, and the 1999 Flora Protection Order in the Republic.
9 Hails C., Loh J. and Goldfinger S. eds (2008) *Living Planet Report 2006*, World Wildlife Fund.
10 Royal Horticultural Society www.rhs.org.uk/research/climate_change 'Gardening in the Global Greenhouse: Planting Opportunities and Challenges'.
11 Rosenzweig C., Karoly D., Vicarelli M., Neofotis P., Wu Q., Casassa G., Menzel A., Root T.L., Estrella N., Seguin B., Tryjanowski P., Liu C., Rawlins S. and Imeson A. (2008) 'Attributing Physical and Biological Impacts to Anthropogenic Climate Change', *Nature* 451, 353-358.
12 Hails C., Loh J. and Goldfinger S. eds (2008) *Living Planet Report 2006*, World Wildlife Fund.
13 Dempsey E. and O'Cleary M. (2002) *The Complete Guide to Ireland's Birds*, 2nd Edition, Dublin, Gill and Macmillan.
14 Cunningham P. (2008) *Ireland's Burning: How Climate Change Will Affect You*, Dublin, Poolbeg Press.
15 BBC Newsline, July 2008.
16 Cunningham P. (2008) *Ireland's Burning: How Climate Change Will Affect You*, Dublin, Poolbeg Press.
17 Bob Davidson, Senior Scientific Officer, Environment & Heritage Service, DOENI, press release June 1st 2000.
18 O'Keeffe C., Lynn D., Weir G., Fernandez Valverde F. and Roller J. (2008) *The Status of EU Protected Habitats and Species in Ireland*, National Parks and Wildlife Service, Department of the Environment, Heritage and Local Government, Government of Ireland.
19 Ibid.

Chapter 9

1 'Climate Change Special Report', *Sunday Times*, 4.11.2007, p.45.
2 Tyndall Centre for Climate change Research, *Briefing on climate change and cities*, Dec 2004, p.14.
3 'Climate Change Special Report', *Sunday Times*, 4.11.2007, p.45.
4 *The Guardian*, 22.1.2008. B&Q has pledged to stop selling the heaters once its current stock is sold.
5 Royal Horticultural Society, www.rhs.org.uk/climate_change/plants.asp.
6 www.sunsmart.com.au
7 Arkell B., Darch G. and McEntee P. (eds) (2007) *Preparing for a Changing Climate in Northern Ireland*, Scottish and Northern Ireland Forum for Environmental Research (SNIFFER).
8 Doggett T. (2008) 'Miles Driven in May drop 3.7%', www.reuters.com 28.7.2008.
9 Behan J. and McQuinn K (2002) 'Projections of Greenhouse Gas Emissions from Irish Agriculture', *Outlook 2002: Medium term analysis for the agri-food sector*, Teagasc: Rural Economy Research Centre, Dublin, p.77-86.
10 Anon (2008) 'Kangaroo Farming would help cut Emissions', *Irish Times*, 9.8.2008
11 Sweeney K., Fealy R., McElwain L., Siggins L. and Sweeney J. (2008) *Changing Shades of Green: The Environmental and Cultural Impacts of Climate Change in Ireland*, Irish American Climate Project, Berkeley, USA.
12 Ibid.
13 Sweeney J., Donnelly A., McElwain L. and Jones M. (2002) *Climate Change Indicators for Ireland*, Wexford, Environmental Protection Agency.

Chapter 10

1 Anon (2008) 'Reviewing Climate Change in Northern Ireland'; Press releases, Department of the Environment, HM Government, 8.12.2004.
2 Secretary of State for the Environment, Food and Rural Affairs (2006) *Climate Change: The UK Programme 2006*, HM Government.
3 Ibid.
4 Anon (2008) G8 'endorses Halving Emissions by 2050', *Irish Times,* Breaking News 8.7.2008.
5 www.wwf.org.uk/core/about/nireland_0000001503.asp.
6 Anon (2008) 'World's First as £12m Turbine Installed in Strangford Lough', *Belfast Telegraph,* 31.3.2008.
7 Burke-Kennedy E. (2008) 'Ireland Set to Miss Kyoto Emissions Target by 10%', *Irish Times*, 27.11.2007.
8 Sammy Wilson, writing in the *Belfast News Letter,* 5.9.2008

9 Ibid.
10 It has recently emerged that Ireland's transport and agriculture emissions may have been underestimated by as much as 12%, meaning that we are further away from the Kyoto target than we thought.
11 Arkell B., Darch G. and McEntee P. (eds) (2007) *Preparing for Climate Change in Northern Ireland*, Scottish and Northern Irish Forum for Environmental Research, UKCC13.
12 Gibbons J. (2008) 'Devastating Climate Change not some Far-off Spectre', *Irish Times*, 21.8.2008.
13 McGee H. (2008) 'Government Advances Climate Change Plan due to Floods', *Irish* Times, 19.8.2008
14 Gibbons J. (2008) 'Devastating Climate Change not some Far-off Spectre', *Irish Times*, 21.8.2008.

Chapter 11
1 Anon (2008) 'World's First as £12m Turbine Installed in Strangford Lough', *Belfast Telegraph,* 31.3.2008.
2 Moriarty G. (2008) 'Feasibility Study on Wind and Wave Energy to involve North and Scotland', *Irish Times,* 8.7.2008.
3 McInerney S. (2008) 'Irish Seaweed may Turn Tide on Oil Crisis', *Sunday Times,* 6.7.2008.
4 www.culgreen.ie
5 *The Ulster Graduate*, 29, Autumn 2008, p.3.
6 Department of the Environment, Food and Rural Affairs (2006) *Climate Change: The UK Programme 2006*, HM Government.

Chapter 12
1 www.sustainablenergyireland.ie
2 www.culgreen.ie
3 These calculations are made by carrying out an energy audit on each listed activity. The audit is conducted by engineers who specialise in energy conservation.
4 Barclay, L. & Grosvenor, M., *Green living for Dummies*, (John Wiley & Sons) p.253.
5 www.combatclimatechange.ie
6 www.sustainablenergyireland.ie
7 www.culgreen.ie
8 www.combatclimatechange.ie

Chapter 13

1 Intergovernmental Panel on Climate Change (IPCC) (2001) *3rd Assessment Report.*
2 Intergovernmental Panel on Climate Change (IPCC) (2007) *4th Assessment Report*, and the Working Group 1 Report 'The Physical Science Basis'.
3 Smithson P., Addison K. and Atkinson K. (2008) *Fundamentals of the Physical Environment*, 4th Edition, Oxford, Routledge.
4 Intergovernmental Panel on Climate Change (IPCC) (2007) *4th Assessment Report,* and the Working Group 1 Report 'The Physical Science Basis'.
5 'Climate Change Special Report', *Sunday Times*, 4.11.2007, p.46-50.
6 Ibid.
7 Data from the US National Snow and Ice Center (NSIDC), June 2008.
8 Communique from International Research Center, Fairbanks, Alaska. Reported AFP 22.9.2008.
9 'A global sea-level rise in excess 44cm can not be discounted'. Pfeffer et al. 2008, *Science*, 321, 1340-1343 suggest that ice melt could add up to two metres to global sea-level.
10 Smithson P., Addison K. and Atkinson K. (2008) *Fundamentals of the Physical Environment*, 4th Edition, Oxford, Routledge.
11 Dr Tony Haymet, CSIRO Marine Research, Australia.
12 Burroughs W.J. (2007) *Climate Change: A Multidisciplinary Approach*, Cambridge University Press.
13 McGuire B. (2002) *A Guide to the End of the World*, Oxford University Press.
14 www.news.mongaybay.com/2006/1012-ucla.

Select bibliography

Aguado, Edward & Burt James E. *Understanding Weather and Climate*, 4th Edition (New Jersey, 2007)

Bell, Martin and Walker, Michael J.C. *Late Quaternary Environmental Change: Physical and Human Perspectives*, 2nd Edition (Essex, 2005)

Bradley, R.S. and Jones, P.D., (eds), *Climate since A.D. 1500*, Revised edition, (London, 1995)

Burroughs William J. *Climate Change: A Multidisciplinary Approach*, 2nd Edition (Cambridge, 2007)

Cabot, David *Ireland: A Natural History* (London, 1999)

Carr, Peter *The Night of the Big Wind*, 2nd Edition, (Belfast, 1993)

– *Portavo: an Irish townland & its peoples*, part one (Belfast, 2003)

Cunningham, Paul *Ireland's Burning: How Climate Change Will Affect You* (Dublin, 2008)

Dawson, A.G., Hickey, K.R., Dawson, S., Elliott, L., Foster, I.D.L., Wadhams, P., Holt, T. Jonsdottir, J., Wilkinson, J., Smith, D.E. and Davis, N. 'Complex North Atlantic Oscillation (NAO) index signal of historic North Atlantic storm track changes', *The Holocene,* vol.12, no.3, p.363-369, 2002

Dawson, A.G., Hickey, K.R., McKenna, J. and Foster, I.D.L. 'A 200-Year Record of Gale Frequency, Edinburgh, Scotland: Possible Links with High-Magnitude Volcanic Eruptions', *The Holocene*, vol.7, No.3, p.337-342, 1997

Dawson, A.G, Hickey, K.R., Mayewski, P. and Nesje, A. 'Greenland (GISP2) ice core and historical indicators of complex North Atlantic climate changes during the fourteenth century', *The Holocene*, vol.17, no.4, p.425-432, 2007

Dickson, David *Arctic Ireland: The Great Frost and Forgotten Famine of 1740-1741* (Belfast 1997)

Douglas Bruce C., Kearney Michael S. and Leatherman Stephen P. *Sea Level Rise: History and Consequences* (San Diego, 2000)

Faidley, Warren *The Ultimate Storm Survival Handbook* (Nashville, 2006)

Flannery, Tim *The Weather Makers: The History and Future Impact of Climate Change* (London, 2005)
Grove, Jean *The Little Ice Age* (London, 1988)
Hardy J.T. *Climate Change: Causes, Effects, Solutions* (Chichester, 2003)
Harvey, L.D. Danny *Global Warming: The Hard Science* (Harlow, 2000)
Hickey, K.R. 'Flooding in the City', in Crowley, J.S., Devoy, R.J.N., Lenihan, D. and O'Flanagan, P. (eds) *Atlas of Cork City*, p.25-31, (Cork, 2005)
Hickey, K.R. 'The Night of the Big Wind: The Impact of the Storm on Loughrea', in Forde, J., Cassidy, C., Manzor, P. and Ryan, D. (eds) *The District of Loughrea: Volume 1 History 1791-1918*, p.130-142, (Loughrea, 2003)
Hickey, K.R. 'The Storminess Record from Armagh Observatory 1796-1999', *Weather*, vol.58, no.1, p.28-35, 2003
Hickey K.R. (2001) 'The Impact of the 31st January 1953 Storm on Scotland', *Scottish Geography Journal*, vol.117, no.4, p.283-295, 2001
Hickey, K.R. and Devoy, R.J.N. 'The Climate of Cork', in Crowley J.S., Devoy
R.J.N., Lenihan, D. and O'Flanagan, P. (eds) *Atlas of Cork City*, p.17-24, (Cork, 2005)
Houghton, John *Global Warming: The Complete Briefing*, 3rd Edition (Cambridge, 2004)
Knight, Jasper (ed) *Field Guide to the Coastal Environments of Northern Ireland* (Coleraine, 2002)
Lynas, Mark *High Tide: How Climate Change is Engulfing our Planet* (London, 2004)
McGuire, Bill *A Guide to the End of the World* (Oxford, 2002)
Murray, Peter *Our Earth: Global Warming The Evidence* (Connecticut, 2007)
Nott Jonathan *Extreme Events: A Physical Reconstruction and Risk Assessment* (Cambridge, 2006)
O'Hare, Greg, Sweeney, John and Wilby, Rob *Weather, Climate and Climate Change: Human Perspectives* (Essex, 2005)
Reynolds, Ross *Weather Rage* (London, 2003)
Rohan, P.K. *The climate of Ireland*, 2nd Edition, (Dublin, 1986)
Sweeney, John 'A three-century storm climatology for Dublin 1715-2000', *Irish Geography*, vol.33, no.1, p.1-14, 2000
Tyrrell, John *Weather and Warfare: A Climatic History of the 1798 Rebellion*, (Cork, 2001).
Viney, Michael *A Living Island: Ireland's Responsibility to Nature* (Dublin, 2003)

Index

Also available from White Row Press:

Arctic Ireland
The Extraordinary story of the Great Frost & Forgotten Famine of 1740-41
David Dickson

Modern Ireland has experienced not one but two Great Famines. One of these, that of 1845-51, needs no introduction. Its place in history is secure.

The other, that of 1740-41, was more intense, more bizarre and proportionately more deadly, yet its existence has been all but forgotten. Arctic Ireland lifts the veil on this older, more enigmatic famine, and attempts to restore it to its rightful place in our historical self-understanding.

On the last day of 1739, Ireland awoke to find itself in the grip of a mini Ice Age.

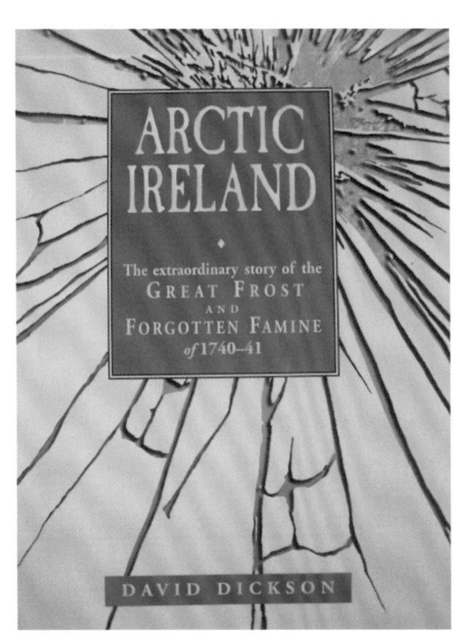

Rivers froze, mills seized up, and houses could not be heated above freezing point. Some people were enchanted by the novelty of it all. Carnivals, dances and sheep-roastings were held on the ice.

But the euphoria proved fleeting. In its wake came an almost biblical ordeal by drought, flood, fire, famine and plague, that has few parallels in the recorded history of the island.

ISBN 1870132858, 96pp, illustrated, £4.95

'The subtitle of this book is "The extraordinary story of the Great Frost & forgotten famine of 1740-41" and for once the adjectives are fully justifed. For the story... is truly extraordinary.' *Sunday Tribune*

'an exemplary study of an unknown period of Irish history - well researched and written and beautifully illustrated.' *Fortean Times*

The Night of the Big Wind
The story of the legendary Big Wind of 1839
Peter Carr

The day began well enough... The children were out enjoying the snow. Indoors all was flutter and bustle, for this was Little Christmas, and everyone was looking forward to the evening's festivities.

At about three o'clock in the afternoon, however, it became almost unnaturally calm - so calm that voices floated between farmhouses more than a mile apart. The temperature soared, until by evening the heat had become sickly. Something strange was happening.

No-one knew exactly what. Maybe it was just as well.

For what followed was a nightmare. What followed was the most terrifying night of their lives...

ISBN 1870132505, 160pp, illustrated, £4.95

'within a few pages you become hooked as if it were a thriller you were reading' *Lisburn Echo*

'An enthralling, gripping read' *Belfast News Letter*

view our books at www.whiterowpress.com